Lecture Notes in Mathematics

T0238079

Editors:
J.-M. Morel, Cachan
F. Takens, Groningen
B. Teissier, Paris

András Telcs

The Art of
Random Walks

Springer

Author

András Telcs
Department of Computer Science
and Information Theory
Budapest University of Technology
Electrical Engineering and Informatics
Magyar tudósok körútja 2,
1117 Budapest
Hungary
e-mail: telcs@szit.bme.hu

Library of Congress Control Number: 2006922866

Mathematics Subject Classification (2000): 60J10, 60J45, 35K05

ISSN print edition: 0075-8434
ISSN electronic edition: 1617-9692
ISBN-10 3-540-33027-5 Springer Berlin Heidelberg New York
ISBN-13 978-3-540-33027-1 Springer Berlin Heidelberg New York

DOI 10.1007/b134090

Springer is a part of Springer Science+Business Media
springer.com
© Springer-Verlag Berlin Heidelberg 2006

Typesetting: by the authors and SPI Publisher Services using a Springer LATEX package
Cover design: *design & production* GmbH, Heidelberg

Printed on acid-free paper   SPIN: 11688020   VA41/3100/SPI   5 4 3 2 1 0

Contents

1

Introduction

1.1 The beginnings

The history of random walks goes back to two classical scientific recognitions. In 1827 Robert Brown, the English botanist published his observation about the irregular movement of small pollen grains in a liquid under his microscope. He not only described the irregular movement but also pointed out that it was caused by some inanimate property of Nature. The irregular and odd series produced by gambling, e.g., while tossing a coin or throwing a dice raised the interest of the mathematicians Pascal, Fermat and Bernoulli as early as in the mid–16th century. Let us start with the physical motivation and then let us recall some milestones in the history of the research on random walks.

The first rigorous results on Brownian motion were given by Einstein [33]. Among other things, he proved that the mean displacement $< X_t >$ of the motion X_t after time t is

$$< X_t >= \sqrt{2Dt},$$

where D is the so–called diffusion constant. Einstein also determined the dependence of the diffusion constant on other physical parameters of the liquid, namely he showed that

$$D^{-1} = \frac{1}{RT} NS$$

where S is the resistance due to viscosity, N is the number of molecules in a unit volume, T is the temperature and $R = 8.3 \times 10^{-7}$ is the gas constant. These results have universal importance. For over half of a century our ideas about diffusion were determined by these laws.

The most natural model of diffusion seems to be the simple symmetric random walk on the $d-$dimensional integer lattice, on $\mathbb{Z}^d$. In this model the moving particle, the (random) walker lives on the vertex set $\mathbb{Z}^d$ and makes steps of unit length in axial directions with probability $P(x, y) = \frac{1}{2d}$. The process is described in discrete time, steps are made at every unit of time.

This classical model is an inexhaustible source of beautiful questions and observations that are useful for sciences, such as physics, economy and biology. It is natural to ask the following questions:

1. How far does the walker get in n steps?
2. How long does it take to cover the a distance R from the starting point?
3. Does the walker return to the starting point?
4. What is the probability of returning?
5. What is the probability of returning in n steps?
6. What is the probability of reaching a given point in n steps?

These questions are the starting points of a number of studies of random walks. There are numerous generalizations of the classical random walk. The space where the random walk takes place may be replaced by other objects like trees, graphs of group automorphism, weighted graphs as well as their random counterparts.

For a long period of time all the results were based on models in which the answer to the first question remained the same; in n steps the walk typically covers $\sqrt{n}$ distance. Let us omit here the exciting subfield of random walks in random environments where other scaling functions have been found (cf. [58],[82]). The answer to the second question is very instructive in the case of simple symmetric random walk on $\mathbb{Z}^d$. Starting at a given point, it takes in average R^2 steps to leave a ball centered at the starting point with radius R. We adopt from physics literature the phrase that in such models we have the $R \to R^2$ space–time scaling.

The diffusion in continuous space is described by the heat diffusion equation

$$\frac{\partial}{\partial t} u = \Delta u, \tag{1.1}$$

which has the following discrete counterpart for random walks:

$$P u_n (x) = \sum_y P(x, y) u_n (y) = u_{n+1} (x),$$

where P is the one step transition operator of the walk. Of course, this can be rewritten by introducing the Laplace operator of random walks

$$\Delta = P - I$$

and the difference operator in time

$$\partial_n u = u_{n+1} - u_n.$$

Using this notation we have the discrete heat equation:

$$\partial_n u = \Delta u. \tag{1.2}$$

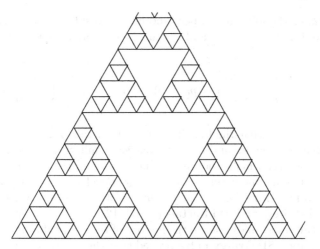

Fig. 1.1. The Sierpinski triangle

The minimal solution of the heat equation on $\mathbb{R}^d$ is given by the classical Gauss-Weierstrass formula

$$p_t\left(x,y\right) = \frac{C_d}{t^{d/2}} \exp\left[-\frac{d^2\left(x,y\right)}{4t}\right].\tag{1.3}$$

It was a long-standing belief that the $R \to R^2$ space-time scaling function rules almost all physical transport processes. This law can be observed in the leading term as well as in the exponent in the Gaussian term. We can consider the leading term as the volume of the ball of radius $\sqrt{t}$ in the d−dimensional space. This term is responsible for the long-time behavior of diffusion, since the second term has no effect if $d^2\left(x,y\right) < t$, while it is the dominant factor if $d^2\left(x,y\right) \gg t$.

The birth of the notion of fractals created among many other novelties a new space-time scaling function: $R \to R^\beta$ with $\beta > 2$. The simplest example of a fractal type object is the Sierpinski triangle shown in Figure 1.1. There are many interesting phenomena to be explored on this graph. Here we focus on the sub-diffusive behavior of the simplest symmetric random walk on it. Due to the bottlenecks between the connected larger and larger triangles the walk is slowed down and as the early works indicated ([84],[1]), and Goldstein proved [37], the mean exit time and consequently the space time scaling function is

$$E\left(x,R\right) \simeq R^\beta,$$

where $\beta = \frac{\log 5}{\log 2} > 2$. Almost at the same time several papers were published discussing the behavior of diffusion on fractals. On fractal spaces and fractal like graphs the following upper (and later two-sided) sub-Gaussian estimates were obtained by Barlow and Perkins [13], Kigami [65] and Jones [60]; $(GE_{\alpha,\beta})$:

$$cn^{-\frac{\alpha}{\beta}}\exp\left[-\left(C\frac{d^\beta(x,y)}{n}\right)^{\frac{1}{\beta-1}}\right] \leq p_n(x,y) \leq Cn^{-\frac{\alpha}{\beta}}\exp\left[-\left(c\frac{d^\beta(x,y)}{n}\right)^{\frac{1}{\beta-1}}\right].$$
(1.4)

Let us emphasize here that the first investigated fractals possess very strong local and global symmetries and in particular self-similarities, which make it possible to develop renormalization techniques analog to the Fourier method. Later Kigami, Hambly and several other authors developed the Dirichlet theory of finitely ramified fractals (cf. [65]). On the other hand, very few results are known on infinitely ramified fractals (cf. [7],[12]).

Now we recall some milestones in the history of the study of random walks. In his famous paper [81] György Pólya proved that the simple symmetric random dom walk on $\mathbb{Z}^d$ returns to the starting position with probability 1 if and only if $d \leq 2$. Much later Nash-Williams [78] proved that this recurrence holds on graphs if and only if the corresponding electric network has infinite resistance in the proper sense. That result very well illuminates the strong connection between the behavior of random walks and the underlying graph as an electric network. In the early 60s Spitzer and Kesten (cf. [88]) developed the potential theory of random walks, while Kemény, Snell, and Knapp [64] and Doob [30] developed the potential theory of Markov chains. The application of the potential theory gained a new momentum with the publication of the beautiful book [32] by Doyle and Snell.

Although the potential theory has a well-developed machinery, it was neglected for long to answer questions mentioned above. Papers devoted to the study of diffusion used algebraic, geometric or spectral properties. In the beginning the classical Fourier method was used, which heavily relies on the algebraic structure of the space [88],[56]. Later spectral properties or isoperimetric inequalities were utilized. All these works (except those about random walks in random environment) remained in the realm of the space-time scaling function R^2 and did not capture the sub-diffusive behavior apparent on fractals.

The new investigations of fractals and in particular, Goldstein [37], Barlow and Perkins [13] and Kusuoka [66] (see also [94]) made it clear that instead of a "one parameter" description of the underlying space two independent features together, the volume growth and resistance growth provide an adequate description of random walks. Goldstein proved the analogue of the Einstein relation for the triangular Sierpinski graph

$$\beta = \alpha + \gamma.$$
(1.5)

Here α is the exponent of the scaling function of the volume of balls, β the exponent of the scaling function of the exit time and γ is the exponent of the resistance. The same relation was given in [94] for a large class of graphs.

In the light of the results on fractals it is natural to raise the following question:

What kinds of properties determine diffusion speed and how are the structural properties of the space reflected in the heat kernel estimates?

The main aim of this book is to answer these questions by utilizing the electric network analogue of random walks on graphs. This model helps to answer these questions in reasonable generality by omitting assumptions on graphs of algebraic nature, symmetry or self-similarity. The results presented here generalize classical works in the discrete context of random walks on graphs (Aronson [2], Moser [71],[72], Cheeger, Yau [23], Gromov [37], Varopoulos [101], Davies [30], and others: [34], [57], [27],[31]) and recent ones on fractals (see Barlow, Bass [8], Kigami [65] and references there). The whole book is devoted to random walks, but there is no doubt that the methods and results can be carried over to metric measure spaces equipped with proper Dirichlet forms. Recent studies successfully transfer results obtained in continuous setting to the discrete graph case and vice versa (cf. [9], [50]).

The text is intended to be self-contained and accessible to graduate or Ph.D. students and researchers with some background of probability theory and random walks or Markov chains. Based on this very limited foundation, some very recent developments in the field are presented with all the technical details. That might slow down reading in some places but provide a possibility to pursue further studies, which is also supported by the inclusion of open problems.

The literature on random walks and diffusion and their applications is so huge that it is hopeless to provide even a partial review of it. The interested reader can find references to start with in Huges' [58],[59] or Woess [107] monographs. Some topic-specific reading is listed below.

- R^n and Riemannian manifolds [26],[40],[44],[55],[93]
- Markov chains and graphs [27],[25],[26],[107]
- random walks on groups and graphs [103],[107],[80]
- fractals [1],[12],[6],[53],[65],[77],[104],
- isoperimetric inequalities, geometry, spectra [20],[24],[68], [74],[75],[79]
- Dirichlet spaces (or measure metric spaces) [7],[35],[57],[65], [82],[90], [89],[105]
- further generalizations [22],[106]

Acknowledgments

Thanks are due to all who understand this exciting field and gave support and inspiration to create my tiny contribution. Among them I am very grateful to Martin Barlow, Richard Bass, Thierry Coulhon, Gábor Elek, Frank den Hollander, András Krámli, Domokos Szász, Bálint Tóth and last but not least to László Györfi. More thanks go to my wife and my kids, András Schubert and others who do not understand even a word of it but gave me support and encouragement to devote myself to this esoteric engagement. I am also grateful to the London Mathematical Society and EPSRC Grant for the financial support. Barry Hughes provided many helpful comments and remarks to improve the draft of this text, many thanks for his invaluable assistance. I am deeply indebted to Alexander Grigor'yan for his friendly support, encouragement and for our enlightening discussions.

Basic definitions and preliminaries

Let us consider a countable infinite connected graph Γ. For sake of simplicity, we assume that there are no multiple edges and loops but with some care almost all of the discussed arguments remain valid in their presence.

A symmetric weight function $\mu_{x,y} = \mu_{y,x} > 0$ is given on the edges $x \sim y$. This weight induces a measure $\mu(x)$

$$\mu(x) = \sum_{y \sim x} \mu_{x,y},$$

$$\mu(A) = \sum_{y \in A} \mu(y)$$

on the vertex set $A \subset \Gamma$ and defines a reversible Markov chain $X_n \in \Gamma$, i.e., a random walk on a weighted graph (Γ, μ) with transition probabilities

$$P(x,y) = \frac{\mu_{x,y}}{\mu(x)},$$

$$P_n(x,y) = \mathbb{P}(X_n = y | X_0 = x).$$

This chain is reversible with respect to μ, since by definition

$$\mu(x) P(x,y) = \mu(y) P(y,x).$$

The transition "density" or heat kernel for the discrete random walk is defined by

$$p_n(x,y) = \frac{1}{\mu(y)} P_n(x,y),$$

and it is clear that it is symmetric:

$$p_n(x,y) = p_n(y,x)$$

The graph is equipped with the usual (shortest path length) graph distance $d(x,y)$ and open metric balls are defined for $x \in \Gamma$, $R > 0$ as

$$B(x, R) = \{y \in \Gamma : d(x, y) < R\},$$
$$S(x, R) = \{y \in \Gamma : d(x, y) = R\}.$$

In the whole text R, r, S will be non negative integers if other is not stated.

Definition 2.1. *In several statements we assume that condition (p_0) holds, that is, there is a universal $p_0 > 0$ such that for all $x, y \in \Gamma, x \sim y$*

$$\frac{\mu_{x,y}}{\mu(x)} \geq p_0. \tag{2.1}$$

For the reader's convenience, a list of important conditions is provided at the end of the book.

Definition 2.2. *On a weighted graph (Γ, μ) the inner product is considered with respect to the measure.*

$$(f, g) = (f, g)_\mu = \sum_{x \in \Gamma} f(x) g(x) \mu(x).$$

Definition 2.3. *The Green function of the random walk is the sum of the probabilities;*

$$G(x, y) = \sum_{n=0}^{\infty} P_n(x, y),$$

and the Green kernel is defined as

$$g(x, y) = \frac{1}{\mu(y)} G(x, y).$$

Definition 2.4. *The random walk is recurrent if $G(x, y) = \infty$ for some $x, y \in \Gamma$ and transient otherwise.*

It is well known that in our setting if

$$G(x, y) < \infty$$

for a given pair of vertices, then it is true for all of them. For a nice introduction to the type problem (distinguishing transient and recurrent graphs), see [107] or [32].

Definition 2.5. *The transition operator with respect to the one step transition probability is defined as*

$$Pf(x) = \sum_y P(x, y) f(y).$$

Definition 2.6. *For $A \subset \Gamma$, $P^A = P^A(y, z) = P(y, z)|_{A \times A}$ is a sub-stochastic matrix (or the operator corresponding to the Dirichlet boundary condition), it is the restriction of P to the set A. Its iterates are denoted by P_k^A and it defines also a random walk killed at exiting from the set A.*

Definition 2.7. *We introduce*

$$G^A(y, z) = \sum_{k=0}^{\infty} P_k^A(y, z)$$

the local Green function, the Green function of the killed walk and the corresponding Green kernel as

$$g^A(y, z) = \frac{1}{\mu(z)} G^A(y, z).$$

Remark 2.1. For finite sets A, the local Green function G^A is finite. The typical choice is $A = B(x, R)$. To study the heat kernel $p_t(x, y)$, it is very useful to consider the Dirichlet heat kernel $p_t^{B(z,R)}(x, y)$ and $G^{B(z,R)}(x, y)$. The main concern is to verify if the Dirichlet heat kernel and local Green function provide an unbiased picture of the global heat kernel as R goes to infinity. In other words whether the approximation of Γ by finite balls provides a correct picture or not. In some respect that is the main point of the whole book.

Definition 2.8. *Let ∂A denote the boundary of a set $A \subset \Gamma : \partial A = \{z \in A^c : z \sim y \in A\}$. The closure of A will be denoted by $\overline{A}$ and defined by $\overline{A} = A \cup \partial A$, $A^c = \Gamma \backslash A$.*

Definition 2.9. $c_0(A)$ *denotes the set of functions with support in A.*

Definition 2.10. *For two real series $a_\xi, b_\xi, \xi \in S$ we shall use the notation $a_\xi \simeq b_\xi$ if there is a $C \geq 1$. such that for all $\xi \in S$*

$$\frac{1}{C} a_\xi \leq b_\xi \leq C a_\xi.$$

Unimportant constants will be denoted by c and C and their value might change from place to place. Typically $C \geq 1$. and $0 < c < 1$.

2.1 Volume

The μ−measure of the ball

$$B(x, R) = \{y \in \Gamma : d(x, y) < R\}$$

is denoted by $V(x, R)$

$$V(x, R) = \mu(B(x, R)).$$

For convenience we introduce a short notation for the volume of the annulus $B(x, R) \setminus B(x, r), R > r > 0$:

$$v = v(x, r, R) = V(x, R) - V(x, r).$$

Definition 2.11. *A graph has polynomial volume growth* (V_a) *with exponent* $\alpha > 0$ *if for all* $x \in \Gamma$

$$V(x, R) \simeq R^\alpha. \tag{2.2}$$

In the literature there are several other expressions for this growth condition: Ahlfors regularity, the graph has a (Hausdorff or fractal) dimension α etc.

Definition 2.12. *A weighted graph satisfies the* volume comparison principle (**VC**) *if there is a constant* $C_V > 1$ *such that for all* $x \in \Gamma$ *and* $R > 0, y \in B(x, R)$

$$\frac{V(x, 2R)}{V(y, R)} \leq C_V. \tag{2.3}$$

Definition 2.13. *A weighted graph has the* volume doubling *property* (**VD**), *if there is a constant* $D_V > 0$ *such that for all* $x \in \Gamma$ *and* $R > 0$

$$V(x, 2R) \leq D_V V(x, R). \tag{2.4}$$

Definition 2.14. *The* weak volume comparison principle (**wVC**) *holds if there is a* $C > 0$ *such that for all* $x \in \Gamma, R > 0, y \in B(x, R)$

$$\frac{V(x, R)}{V(y, R)} \leq C, \tag{2.5}$$

Definition 2.15. *The anti-doubling condition for the volume* (**aDV**) *holds if there is an* A_V *such that for all* $x \in \Gamma, R > 0$

$$2V(x, R) \leq V(x, A_V R). \tag{2.6}$$

Remark 2.2. It is evident that (VC) and (VD) are equivalent.

Lemma 2.1. *If (p_0) and (VD) hold, then*
1. for all $x \in \Gamma, R > 0$, $y \in B(x,R)$ the weak volume comparison condition (wVC) holds,
2. the anti-doubling condition for the volume (aDV) holds,
3. for all $x \in \Gamma, R \geq 1$

$$V(x, MR) - V(x, R) \simeq V(x, R) \tag{2.7}$$

for any fixed $M \geq 2$.

Proof   The validity of (wVC) and (2.7) is evident, (aDV) can be seen following the proof of [27, Lemma 2.2]. Since Γ is infinite and connected there is a $y \in \Gamma, d(x, y) = 3R$ and (VC) implies that there is a fixed $\varepsilon > 0$ such that

$$V(y, R) \geq \varepsilon V(x, R).$$

This yields that

$$V(x, 4R) \geq V(x, R) + V(y, R) \geq (1 + \varepsilon) V(x, R).$$

Iterating this inequality enough times (aDV) follows.   ∎

Remark 2.3. As we already mentioned (VC), (cf. [27, Lemma 2.2]) is equivalent to (VD) and it is again evident that both are equivalent to the inequality

$$\frac{V(x, R)}{V(y, r)} \leq C \left(\frac{R}{r} \right)^{\alpha}, \tag{2.8}$$

where $\alpha = \log_2 C_V$ and $d(x, y) < R, R > r > 0$, which is the original form of Gromov's volume comparison inequality (cf. [51]).

Remark 2.4. The anti-doubling property has the following equivalent form. There are $c, \alpha' > 0$ such that for all $x \in \Gamma, R > r$

$$c \left(\frac{R}{r} \right)^{\alpha'} \leq \frac{V(x, R)}{V(y, r)}. \tag{2.9}$$

Thanks to inequality (2.7) in the sequel we can use

$$v(x, R, 2R) = V(x, 2R) - V(x, R) \simeq V(x, R)$$

provided that (p_0) and (VD) hold.

Proposition 2.1. *If (p_0) holds, then for all $x, y \in \Gamma$ and $R > 0$ and for some $C > 1$,*

$$V(x, R) \leq C^R \mu(x), \tag{2.10}$$

$$p_0^{d(x,y)} \mu(y) \leq \mu(x) \tag{2.11}$$

and for any $x \in \Gamma$

$$|\{y : y \sim x\}| \leq \frac{1}{p_0}. \tag{2.12}$$

Proof   Let $x \sim y$. Since $P(x,y) = \frac{\mu_{xy}}{\mu(x)}$ and $\mu_{xy} \leq \mu(y)$, the hypothesis (p_0) implies $p_0 \mu(x) \leq \mu(y)$. By symmetry, we also have

$$p_0 \mu(y) \leq \mu(x).$$

Iterating these inequalities we obtain, for arbitrary x and y,

$$p_0^{d(x,y)} \mu(y) \leq \mu(x). \tag{2.13}$$

and this is (2.11). Therefore, any ball $B(x,R)$ has at most C^R vertices inside. By (2.13), any point $y \in B(x,R)$ has measure at most $p_0^{-R} \mu(x)$, whence (2.10) follows.   ∎

Definition 2.16. *The* bounded covering principle *(***BC***) holds if there is a fixed K such that for all $x \in \Gamma$, $R > 0$, $B(x, 2R)$ can be covered with at most K balls of radius R.*

It is well known that the volume doubling property implies the bounded covering principle.

Proposition 2.2. *If (p_0) and (VD) hold, then (BC) holds as well.*

Proof   Assume first that $R \geq 2$. Consider $B(x, R)$ and the maximal possible packing of it with (non-intersecting) balls of radius $R/2$. From Proposition 2.1 we know that we have a finite number of packing balls, denote this number by K. First we show that $\{B(x_i, R), i = 1..K\}$ is a covering. Assume that the centers are $x_1, x_2 ... x_K$, $i = 1, ..., K$. If there is an uncovered vertex $z \in B(x, R)$, i.e. , for all $i = 1, ...K$

$$d(z, x_i) \geq R,$$

then $B(z, R/2)$ can be added to the packing since it has no intersection with any $B(x_i, R/2)$, which contradicts to maximality. On the other hand, it is clear that

$$\bigcup_{i=1}^{K} B(x_i, R/2) \subset B(x, 2R),$$

and from the non-intersection of $B(x_i, R/2)$s that

$$\sum_{i=1}^{K} V(x_i, R/2) \leq V(x, 2R).$$

Finally by using (2.8), the consequence of (VD), we obtain

$$K \min V(x_i, R/2) \leq \sum_{i=1}^{K} V(x_i, R/2) \leq V(x, 2R)$$

$$K \leq \frac{V(x, 2R)}{\min_i V(x_i, R/2)} \leq \max_i \left\{ \frac{V(x, 2R)}{V(x_i, R/2)} \right\} \leq C4^{\alpha}.$$

If $R < 2$, then the statement follows from (p_0) and Proposition 2.1.   ∎

Remark 2.5. We can easily see that

$$(wVC) + (BC) \Longleftrightarrow (VD).$$

Exercise 2.1. Show the above implication.

2.2 Mean exit time

Let us introduce the exit time T_A for a set $A \subset \Gamma$.

Definition 2.17. *The exit time from a set A is defined as*

$$T_A = \min\{k \geq 0 : X_k \in A^c\},$$

its expected value is denoted by

$$E_x(A) = \mathbb{E}(T_A | X_0 = x),$$

and we will use the short notations, $T_{x,R} = T_{B(x,R)}$, $E = E(x,R) = E_x(x,R) = E_x(B(x,R))$.

Definition 2.18. *The hitting time of a set A is*

$$\tau_A = T_{A^c}.$$

Remark 2.6. Let us observe that the exit time can be expressed with the Green function (cf. Definition 2.7):

$$E_x(A) = \sum_{y \in A} G^A(x,y) = \sum_{y \in A} \sum_{i=0}^{\infty} P_i^A(x,y).$$

Definition 2.19. *The maximal mean exit time is defined as*

$$\overline{E}(A) = \max_{x \in A} E_x(A)$$

and in particular the simplified notation $\overline{E}(x,R) = \overline{E}(B(x,R))$ will be used.

Definition 2.20. *The graph (Γ, μ) has the property $(\overline{\mathbf{E}})$ if there is a $C \geq 1$. such that for all $x \in \Gamma$, $R > 1$*

$$\overline{E}(x,R) \leq CE(x,R).$$

Definition 2.21. *The mean exit time is uniform in the space if for all $x, y \in \Gamma$*

$$E(x,R) \simeq E(y,R).$$

This condition will be referred to by $(\mathbf{E})$.

Definition 2.22. *We will say that a weighted graph* (Γ, μ) *satisfies the time comparison principle* (**TC**) *if there is a constant* $C \geq 1$. *such that for all* $x \in \Gamma$ *and* $R > 0, y \in B(x, R)$

$$\frac{E(y, 2R)}{E(x, R)} \leq C. \tag{2.14}$$

Definition 2.23. *We will say that a weighted graph* (Γ, μ) *satisfies the weak time comparison principle* (**wTC**) *if there is a constant* $C \geq 1$. *such that for all* $x \in \Gamma$ *and* $R > 0, y \in B(x, R)$

$$\frac{E(x, R)}{E(y, R)} \leq C. \tag{2.15}$$

Definition 2.24. *We will say that a weighted graph* (Γ, μ) *satisfies the time doubling property* (**TD**) *if there is a constant* $C_T > 1$ *such that for all* $x \in \Gamma$ *and* $R > 0$

$$\frac{E(x, 2R)}{E(x, R)} \leq C_T. \tag{2.16}$$

Exercise 2.2. Prove that $(TC) \Longleftrightarrow (TD) + (wTC)$.

2.3 Laplace operator

The Laplace operator plays a central role in the study of random walks. The full analogy between diffusion on continuous spaces and random walks on weighted graphs extends to the Laplace operator.

Definition 2.25. *The Laplace operator on a weighted graph* (Γ, μ) *is defined simply as*

$$\Delta = P - I.$$

Definition 2.26. *The Laplace operator with Dirichlet boundary conditions on finite sets, in particular for balls, can be defined as*

$$\Delta^A f(x) = \begin{cases} \Delta f(x) & \text{if } x \in A \\ 0 & \text{if } x \notin A \end{cases},$$

for $f \in c_0(A)$. *The smallest eigenvalue of* $-\Delta^A$ *is denoted in general by* $\lambda(A)$ *and for* $A = B(x, R)$ *by* $\lambda = \lambda(x, R) = \lambda(B(x, R))$.

Remark 2.7. From the Perron-Frobenius theorem we know that all the eigenvalues of $P|_{A \times A}$ are real and the largest one is $0 < 1 - \lambda(A) < 1$ and it has multiplicity one.

Let us introduce the differentia operator along the edges (x, y) :

$$\nabla_{x,y} f = f(y) - f(x),$$

and observe that

$$\Delta f = P\nabla f$$
$$P\nabla f(x) = \sum_{x,y} \frac{1}{\mu(x)} \left(\nabla_{x,y} f\right) \mu_{x,y},$$

and the discrete Green formula can be given as follows:

$$(\Delta f, g) = \sum_{x} (\Delta f)(x) g(x) \mu(x) = -\frac{1}{2} \sum_{x,y} \left(\nabla_{x,y} f\right) \left(\nabla_{x,y} g\right) \mu_{x,y}. \qquad (2.17)$$

Definition 2.27. *The energy or Dirichlet form $\mathcal{E}(f, f)$ associated with an electric network can be defined via the bilinear form*

$$(\Delta f, g) = -\frac{1}{2} \sum_{x,y} \left(\nabla_{x,y} f\right) \left(\nabla_{x,y} g\right) \mu_{x,y}$$

as

$$\mathcal{E}(f, f) = -(\Delta f, f) = \frac{1}{2} \sum_{x,y \in \Gamma} \mu_{x,y} \left(f(x) - f(y)\right)^2.$$

Using this notation the smallest eigenvalue of $-\Delta^A$ can be defined by

$$\lambda(A) = \inf\left\{\frac{\mathcal{E}(f, f)}{(f, f)} : f \in c_0(A)\right\} \qquad (2.18)$$

as well.

Lemma 2.2. *If (Γ, μ) satisfies the volume doubling property (VD), then for all $x \in \Gamma$ and $R > 0$,*

$$\lambda(x, R) \le CR^{-2} \qquad (2.19)$$

Proof   Let us apply the variational definition (2.18) with the test function

$$f(y) = (R - d(x, y))_+ \in c_0(B(x, R)).$$

Since $|\nabla_{yz} f| \le 1$ and (2.18) holds (VD) imply

$$\lambda(x, R) \le \frac{\frac{1}{2} \sum_{y \sim z} (\nabla_{yz} f)^2 \mu_{yz}}{\sum_y f^2(y) \mu(y)} \le \frac{CV(x, R)}{R^2 V(x, R/2)} \le C' R^{-2},$$

the statement follows.. ∎

Lemma 2.3. *For all (Γ, μ) and for all finite $A \subset \Gamma$*

$$\lambda^{-1}(A) \leq \overline{E}(A). \tag{2.20}$$

Proof   Let us assume that $f \geq 0$ is the eigenfunction corresponding to $\lambda = \lambda(A)$, the smallest eigenvalue of the operator $-\Delta^A = I - P^A$ on A and let us normalize f so that $max_{y \in A} f(y) = f(x) = 1$. It is clear that

$$E_x(T_A) = \sum_{y \in A} G^A(x, y),$$

while $(-\Delta_A)^{-1} = G^A$, consequently

$$\frac{1}{\lambda} = \frac{1}{\lambda} f(x) = \left(G^A f \right)(x) \leq \sum_{y \in A} G^A(x, y) = E_x(T_A) \leq \max_{z \in A} E_z(T_A),$$

which gives the statement.   ∎

2.4 Resistance

Definition 2.28. *For any two disjoint sets, $A, B \subset \Gamma$, the resistance $\rho(A, B)$, is defined as*

$$\rho(A, B) = (\inf \{ \mathcal{E}(f, f) : f|_A = 1, f|_B = 0 \})^{-1} \tag{2.21}$$

and we introduce

$$\rho(x, r, R) = \rho(B(x, r), \partial B(x, R))$$

for the resistance of the annulus about $x \in \Gamma$, with radii $R > r \geq 0$.

The formal definition of the resistance is in full agreement with the physical interpretation and our naive understanding. We can consider the edges as conductors with conductance or capacity $\mu_{x,y}$ or resistors with resistance $1/\mu_{x,y}$. The whole graph can be considered as an electric network with wires represented by the edges. The effective resistance $\rho(A, B)$ is the voltage needed to produce unit current between the sets A, B if they are connected to a potential source. Let us recall that Ohm's law says that $\rho = \frac{U}{I}$, where U is the potential difference between the two ends (terminals) of the electric network and I is the resulting current. It is easy to see (cf. [64]) that all reversible Markov chains have such electric network interpretations and vice versa any electric network (containing resistors only) determines a reversible Markov chain. Some further discussion on the electric network model will be given in Chapter 3.

Definition 2.29. *We shall say that the annulus resistance is uniform in the space relative to the volume if there is an $M > 1, M \in \mathbb{N}$ such that for all $x, y \in \Gamma, R \geq 0$ ($\rho\mathbf{v}$) holds:*

$$\rho(x, R, MR)v(x, R, MR) \simeq \rho(y, R, MR)v(y, R, MR). \qquad (2.22)$$

In some of the proofs we shall use the refined wire model where edges are considered as homogeneous resistors of conductance $\mu_{x,y}$ (see [94]). Let us identify edges by unit intervals and assume that resistances are proportional to the length. In this way a continuous measure metric space is in our possession. Harmonic functions, in particular, Green functions extend linearly along the wires.

Definition 2.30. *Let $\overrightarrow{\Gamma} = \{\overrightarrow{(x, y)} : x, y \in \Gamma, x \sim y\}$ denote the set of (arbitrarily) oriented edges of Γ. $W = \overrightarrow{\Gamma} \times [0, 1]$ and for a function h and for an $x \in \Gamma$ the equipotential surface of x is*

$$\Gamma_x = \{w = \overrightarrow{(y, z)} \times \{\alpha\} \in W : (1 - \alpha) h(y) + \alpha h(z) = h(x)\},$$

and similarly, $\Gamma_w = \{s \in W : h(s) = h(w)\}$ for a $w \in W$.

This setup allows us to speak about equipotential surfaces, and these surfaces are the boundaries of super-level sets of harmonic functions, which will be introduced later. At some given points we shall refer to the objects of the refined model using the notation $(.)^r$, e.g., as to a set H by H^r.

Definition 2.31. *The resistance between a finite set A and "infinity" is defined by*

$$\rho(A) = \sup_{B : A \subset B} \rho(A, B^c) . \qquad (2.23)$$

For $A = \{x\}$ we use the shorter notation $\rho(x) = \rho(\{x\})$.

Several criteria are known for the transience of random walks. As we mentioned in the introduction the first one was given by Pólya [81] and a very general one was given by Nash-Williams [78]. Let us recall here a combined statement, which helps to capture the connection between some of the objects defined above and transience. (cf. [107] or [32])

Theorem 2.1. *For connected, infinite, weighted graphs, the following statements are equivalent.*

1. *A random walk is transient.*
2. *For any $f \in c_0(\Gamma)$, the energy form $\mathcal{E}(f, f)$ is finite.*
3. *There is an $x \in \Gamma$ such that $\rho(x) < \infty$.*

2.5 Model fractals

In this section we define and study the simplest fractal-like graphs, which will serve as recurring examples in the whole sequel. The Sierpinski triangular graph or the pre-Sierpinski gasket SG_2 is the test-bed for almost all studies of diffusion on fractals ([84],[1],[37] just to recall some early works). The pre-Sierpinski gasket is an infinite graph which is defined in a recursive way. Let S_0 be a triangle graph, $S_0 = \{a, b, c\}$. Consider three copies of it S_a, S_b, S_c, where the new S_i sets are $S_i = \{ai, bi, ci\}$ and for all possible $i, j = a, b, c$ we identify the vertices $ij = ji$. In this way we have obtained the level one triangle S_1. Let us repeat the same procedure with S_1 and so forth. The starting chunk of the graph is shown in Figure 1.1. Some properties of the graph are straightforward, some other ones need some studying. We state them without proof, which is left to the reader as an exercise.

Proposition 2.3. *The volume growth of* SG_2 *satisfies*

$$V(x, R) \simeq R^{\alpha},$$

where $\alpha = \frac{\log 3}{\log 2}$. *If all the edges represent unit resistors, the resistance satisfies*

$$\rho(x, R, 2R) \simeq R^{\gamma},$$

where $\gamma = \frac{\log 5 - \log 3}{\log 2}$, *finally for the mean exit time we have*

$$E(x, R) \simeq R^{\beta},$$

where $\beta = \frac{\log 5}{\log 2} = \alpha + \gamma$.

That means that on this graph all the basic quantities grow polynomially and since $\beta > 2$, the random walk is sub-diffusive on it. This behavior is produced by the tiny links between the larger and larger triangles. It is difficult for the random walk to transfer from one large triangle to another one and this problem repeats itself on all levels. All the above asymptotic relations can be derived by using recurrence relations based on the nice recursive structure of the graph. The two-sided sub-Gaussian estimate $(GE_{\alpha,\beta})$ is proved by Jones [60], for further results and generalizations see [9],[11],[65].

Let us recall a well-known method to form a Riemannian manifold based on a graph.

As Figure 2.1 shows we replace the vertices by balls and the edges of the graph with tubes which join smoothly with the balls. It is apparent that heat diffusion on the surface of the balls and tubes behaves locally as in $\mathbb{R}^2$ but globally it does not, rather it moves similarly to a random walk on the discrete structure.

The converse procedure is also known. Given a measure metric space, we can construct its ε-nett, which is a graph and a random walk on this graph approximates the continuous heat propagation in the space (cf. [9]).

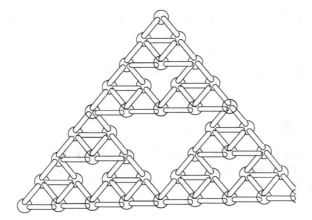

Fig. 2.1. The Sierpinski gym

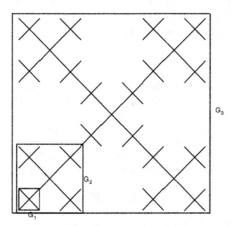

Fig. 2.2. The Vicsek tree

The second and even more elementary graph which possesses fractal-type properties is the Vicsek tree. Let G be the Vicsek tree (embedded in $\mathbb{R}^2$) – see Figure 2.2 – which is the union of an increasing sequence of blocks $\{G_k\}_{k=1}^{\infty}$. Here $G_0 = \{o\}$ and G_{k+1} consists of G_k and its four copies are translated and glued in an obvious way. The basic properties of the Vicsek tree can be easily seen.

Proposition 2.4. *The volume growth satisfies*

$$V(x, R) \simeq R^{\alpha},$$

where $\alpha = \frac{\log 5}{\log 3}$. *If all the edges represent unit resistors, the resistance satisfies*

$$\rho\left(x, R, 2R\right) \simeq R,$$

so $\gamma = 1$ *and finally for the mean exit time we have*

$$E\left(x, R\right) \simeq R^{\beta},$$

where $\beta = 1 + \frac{\log 5}{\log 3} = \alpha + \gamma$.

Later we shall deduce $(GE_{\alpha,\beta})$ and its generalizations for the Vicsek tree. Let us remark that this tree contains only one infinite path (the backbone in Kesten's terminology) and the increasing dead-ends slow down the random walk.

Example 2.1. Here we describe a graph in which $V(x, R)$ substantially depends on x. Let G be the Vicsek tree (embedded in $\mathbb{R}^2$) which is the union of an increasing sequence of blocks $\{G_k\}_{k=1}^{\infty}$.

Fix $Q \geq 1$ and define weight μ_{xy} for any edge $x \sim y$ by $\mu_{xy} = Q^k$ where k is the minimal index so that Γ_k contains x or y. Since $d(x, o) \simeq 3^k$ for any $x \in \Gamma_k \setminus \Gamma_{k-1}$, this implies that for all $x \neq o$

$$\mu(x) \simeq d(x, o)^{\delta}, \tag{2.24}$$

where $\delta = \log_3 Q$. (See Figure 2.3.)

Let x_k be the symmetry center of Γ_k and set $R_k = 3^{k-1} + \frac{1}{2}$, then $\Gamma_k = B(x_k, R_k)$. Clearly,

$$|B(x_k, R_k)| = |\Gamma_k| \simeq 5^k \simeq R_k^{\alpha}$$

where $|\cdot|$ is the cardinality of a set and $\alpha = \log_3 5$. It is not difficult to see that the same relation holds for all balls $B(x, R)$ in Γ with $R \geq 1$, that is

$$|B(x, R)| \simeq R^{\alpha}. \tag{2.25}$$

From (2.24) and (2.25) we easily obtain that for all $x \in \Gamma$ and $R \geq 1$

$$V(x, R) = \mu\left(B(x, R)\right) \simeq R^{\alpha}\left(R + d\left(x, o\right)\right)^{\delta}. \tag{2.26}$$

It is clear that (2.26) holds and consequently (VD) is satisfied.

Due to the tree structure of Γ, it is easy to compute the Green kernel $g_k(y) = g^{\Gamma_k}(x_k, y)$.

It will be shown that

$$E\left(x_k, R_k\right) = \sum_{y \in \Gamma_k} g_k(y) \mu\left(y\right) \simeq 3^k |\Gamma_k| \simeq 15^k \simeq R_k^{\beta}$$

where $\beta = \log_3 15$. It is easy to see that the same relation $E(x, R) \simeq R^{\beta}$ holds for all $x \in \Gamma$ and $R \geq 1$, which proves (E_{β}).

The Green kernel $g_k = g^{\Gamma_k}(x_k, \cdot)$ constructed above is nearly radial. A similar argument shows that the same is true for all balls in Γ.

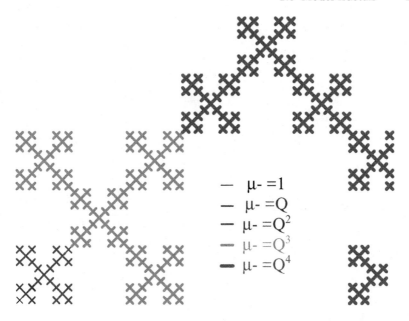

Fig. 2.3. The weighted Vicsek tree

Part I

Potential theory and isoperimetric inequalities

3

Some elements of potential theory

In this chapter first we recall some elementary and nice arguments from [32]. The electric network analogue of reversible Markov chains provides natural interpretation and explanation for many phenomena. We think that this intuitive background helps to understand the presented results. Then the interplay between the elliptic Harnack inequality and the behavior of the Green functions is explained.

3.1 Electric network model

Consider a finite weighted graph (Γ, μ) and two vertices $a \neq b$. We are interested in the probability of reaching a before reaching b if the walk starts at x. Formally $\tau_y = \min \{k : X_k = y\}$ denote the first hitting time of y and

$$h(x) = h_{a,b}(x) = \mathbb{P}(\tau_b > \tau_a | X_0 = x).$$

We can easily observe that for all $x \neq a, b$

$$h(x) = \sum_{y \sim x} P(x, y) h(y).$$

Definition 3.1. *A function $h : \Gamma \to \mathbb{R}$ is harmonic (on the whole Γ) if*

$$\triangle h = 0 \tag{3.1}$$

and harmonic on $A \subset \Gamma$ if it is defined on $\overline{A}$ and

$$\triangle h(x) = 0 \text{ for all } x \in A.$$

It is clear that (3.1) is equivalent to

$$Ph = h \text{ on } A, \tag{3.2}$$

$$\sum_{y \sim x} P(x, y) h(y) = h(x) \quad \text{for } x \in A.$$

Now let us study the voltages $v(x)$ produced by the external potential source keeping $v(a) = 1, v(b) = 0$. Kirchoff's node law says that the sum of the $i_{x,y}$ currents on all $x \neq a, b$ is zero. Using Ohm's law for a fixed $x \in \Gamma \backslash \{a, b\}$ we have

$$0 = \sum_{y \sim x} i_{x,y} = \sum_{y \sim x} \mu_{x,y} (v(y) - v(x)),$$

which can be reformulated as

$$\sum_{y \sim x} \frac{\mu_{x,y}}{\mu(x)} v(y) = v(x).$$

Let us recall that $P(x, y) = \frac{\mu_{x,y}}{\mu(x)}$ and $h(a) = 1, h(b) = 0$ by definition, which makes the two systems of linear equations identical. Since Γ is connected and finite, their solutions are the same:

$$h(x) = \mathbb{P}(\tau_b > \tau_a | X_0 = x) = v(x).$$

A similarly interesting observation can be made if instead of $v(a) = 1$ we assume that $v(a) = \rho(a, b)$, which again by Ohm's law produces exactly a unit of current from a to b. If we rearrange

$$\sum_{y \sim x} \frac{\mu_{x,y}}{\mu(x)} v(y) = v(x)$$

for a fixed $x \in \Gamma$, it follows that

$$\sum_{y \sim x} \mu_{x,y} v(y) = \mu(x) v(x),$$

and by $\mu_{x,y} = \mu_{y,x}$ and the notation $u(x) = \mu(x) v(x)$,

$$\sum_{y \sim x} \frac{\mu_{y,x}}{\mu(y)} \mu(y) v(y) = \mu(x) v(x),$$

$$\sum_{y \sim x} u(y) P(y, x) = u(x).$$

In this way we have a right harmonic function (or more correctly a harmonic measure) with respect to P. It is easy to observe that

$$u(y) = G^{\Gamma \backslash \{b\}}(a, y),$$

the Green's function, or the local time at y (starting at a and killed when hitting b), satisfies the same equation for all $x \neq a, b$

$$\sum_{y \sim x} u(y) P(y, x) = u(x).$$

Now we investigate the probability of reaching b before reentering in a. That means that we allow one initial step from a. This is called escape probability: P_{esc}. The usual renewal argument shows that

$$G^{\Gamma\backslash b}(a,a) = \frac{1}{P_{esc}}, \tag{3.3}$$

and

$$G^{\Gamma\backslash b}(a,a) = \mu(a)\rho(a,b), \tag{3.4}$$

which results in

$$P_{esc} = \frac{1}{\mu(a)\rho(a,b)}. \tag{3.5}$$

This nice analogue between potential values on an the electric network and escape probabilities can be utilized to capture the exit time property as well. It is worth mentioning that

$$E_x(\Gamma\backslash\{b\}) = \sum_{y\in\Gamma} G^{\Gamma\backslash b}(x,y)$$

for all $x \in \Gamma\backslash\{b\}$. That shows that better understanding of the behavior of harmonic functions results in better understanding of the mean exit time. In particular, harmonic functions satisfy the maximum principle, which ensures the existence of potential levels and their shapes contain very useful information.

Exercise 3.1. Show the identity (3.3).

Exercise 3.2. A function h is harmonic on finite set $A \subset \Gamma$ if and only if

$$h(x) = \mathbb{E}_x(h(X_{T_A})).$$

Proposition 3.1. *If h is harmonic in a finite set A, then its maximum and minimum are attained at the boundary.*

Proof   Without loss of generality we can assume that $h \geq 0$ on $\overline{A}$. If h is constant on $\overline{A}$, there is nothing to prove. Assume that the maximum is attained at $x \in \overline{A}$, for which there is an $y \sim x, y \in A$ $h(y) < h(x)$. Such a y must exist, otherwise h is constant. Assume that $x \in A$. Since h is harmonic,

$$h(x) = \sum_{y\sim x} P(x,y)h(y) = P(x,y)h(y) + \sum_{z\sim x, z\neq y} P(x,z)h(z)$$
$$\leq P(x,y)h(y) + [1 - P(x,y)]h(x),$$

which reduces to

$$h(x) \leq h(y)$$

contradicting with our hypothesis. For the minimum consider a large constant $C > h(x)$ and apply the same argument to $u(x) = C - h(x)$.   ∎

From the maximum principle, it is standard to deduce uniqueness.

Corollary 3.1. *If g, h are harmonic on a finite A and $g = h$ on ∂A, then $g = h$ on A as well.*

It is also useful to observe that the mean exit time satisfies the following equation

$$E_x(A) = 1 + \sum_{y \sim x} P(x, y) E_y(A) \qquad (3.6)$$

for $x \in \Gamma$.

Let us recall the definition of effective resistance between two sets.

$$\frac{1}{\rho(A, B)} = \inf \left\{ \sum_{x,y} \mu_{x,y} \left(f(x) - f(y) \right)^2 : f|_A = 1, f|_b = 0 \right\}$$

From this, it is evident that an increase of any $\mu_{x,y}$ cannot result in an increase of the resistance $\rho(A, B)$ and vice versa, the decrease of any $\mu_{x,y}$ cannot result in its decrease. This is the monotonicity principle, which is based on comparison of Dirichlet forms and has useful and practical consequences which help to compare weighted graphs and study random walks on them.

Corollary 3.2. *1 If on Γ we have two weights $\mu \leq \mu'$ and the random walk on (Γ, μ) is transient, then it is transient on (Γ, μ') as well.*
2 If we add a new edge to the graph, the resistance between any two sets does not increase.
3 If we remove an edge from the graph, the resistance between any two sets does not decrease.
4 If we shrink two vertices in one, the resistance between any two sets does not increase.
5 If v is an electric potential on the graph and we shrink two vertices having the same potential value, the potential values will not change anywhere.
6 If v is an electric potential on the graph and we remove an edge (change the weight) between two vertices having the same potential value, the potential values will not change anywhere.

The first four statements follow from the monotonicity principle while 5 and 6 follow from the fact that there is no current between nodes having the same potential value. Of course, this observation applies to harmonic functions on (Γ, μ), too. These operations have particular advantages when determining the type of a graph, that is, finding out if it is recurrent or not (cf. [32]). The use of potential levels helps to calculate the mean exit time and to prove the Einstein relation (cf. [94],[108]) (see Chapter 7).

A very useful alternative definition of resistance can be given by using flux or current. The difference at a point x in the direction $y \sim x$ is defined as

$$I\left(u,(x,y)\right) = \nabla_{x,y}u = \mu_{x,y}\left(u\left(y\right) - u\left(x\right)\right).$$

An edge set C is a cut-set of the graph if there are $\Gamma_1, \Gamma_2 \subset \Gamma$ such that $\Gamma_1 \cap \Gamma_2 = \emptyset$, all the paths from Γ_1 to Γ_2 intersect C. If $C \subset E\left(\Gamma\right)$ is a cut-set, the flux (or current) through C from Γ_1 to Γ_2 is defined as

$$I\left(u,C\right) = \sum_{x \in \Gamma_1, y \in \Gamma_2} I\left(u,(x,y)\right) = \sum_{x \in \Gamma_1, y \in \Gamma_2} \nabla_{x,y}u.$$

Here the order of Γ_is is not important, but all the edges in the sum are "oriented" in the same way. Write $C_a = \{(a,y) : y \sim a\}$

$$\rho\left(a,b\right) = \inf\left\{\mathcal{E}\left(u,u\right) : I\left(h,C_a\right) \geq 1\right\}.$$

It is easy to see that if h is a harmonic function on $\Gamma \setminus \{a,b\}$ and $h\left(b\right) = 0, h\left(a\right) = C > 0$, then the flux

$$I\left(h,C_a\right) = \frac{C}{\rho\left(a,b\right)},$$

and

$$\mathcal{E}\left(h,h\right) = CI\left(h,C_a\right),$$

which means that

$$\mathcal{E}\left(h,h\right) = \rho\left(a,b\right) \tag{3.7}$$

if $h\left(a\right) = \rho\left(a,b\right)$. This harmonic function h is called capacity potential on Γ between a and b.

3.2 Basic inequalities

Now we are giving some inequalities which play an important role in the sequel.

Theorem 3.1. *For all graphs $\left(\Gamma,\mu\right)$ and for all finite sets $A \subset B \subset \Gamma$, the inequality*

$$\lambda(B)\rho(A,\partial B)\mu(A) \leq 1 \tag{3.8}$$

holds, in particular for $R > r > 0$

$$\lambda(x,R)\rho(x,r,R)V(x,R) \leq 1. \tag{3.9}$$

Proof   By using the capacity potential between A and ∂B, the first statement follows from the variational definition of the eigenvalue. τ_D denotes the first hitting time of $D \subset \Gamma$ and

$$v\left(y\right) = P_y\left(\tau_A < \tau_{\partial B}\right).$$

By definition (2.18),

$$\lambda\left(B\right) \le -\frac{\left(\Delta v, v\right)}{\left(v, v\right)} = \frac{\sum \mu_{x,y}\left(v\left(x\right) - v\left(y\right)\right)^2}{\left(v, v\right)},$$

and we know that the unit potential generates $1/\left(\rho\left(A, \partial B\right)\right)$ current between A and B^c (see (3.7)). That means that

$$\lambda\left(B\right) \le \frac{1}{\rho\left(A, \partial B\right)\mu\left(A\right)}.$$

The inequality (3.9) is a direct consequence of (3.8).   ∎

Consider finite sets $\emptyset \ne A \subset B \subset \Gamma$. It will be useful to study a random walk on $B\backslash A$ with a reflecting (Neumann) boundary on A and a killing or Dirichlet boundary on $\Gamma\backslash B$. For this, let us shrink the set A in a new point a and the set $\Gamma\backslash B$ in a new point b. The new weights are introduced in a natural way. If $x, y \in B\backslash A$, $\nu_{x,y} = \mu_{x,y}$. For $y \in \partial A$, $z \in B$

$$\nu_{a,y} = \sum_{x\in A} \mu_{x,y};$$

$$\nu_{z,b} = \sum_{w\in B^c} \nu_{z,w};$$

$$\nu\left(a\right) = \sum_{y} \nu_{a,y};$$

$$\nu\left(b\right) = \sum_{z} \nu_{z,b}.$$

The walk is killed when leaving B. Let us remark that $\tau_b = \tau_{B^c} = T_B$. Let us denote the new graph by Γ_a^b.

Proposition 3.2. *For all random walks on weighted graphs and for finite sets* $\emptyset \ne A \subset B \subset \Gamma$

$$\lambda(B) \le \frac{\rho(A, B^c)\mu(B\backslash A)}{E_a(\tau_b)^2},$$

where τ_b denotes the time of the walk that it spends in $B\backslash A$, so that it is reflected on A and is killed when it leaves B.

Proof   Consider the smallest eigenvalue of the Laplacian on $\left(\Gamma_a^b, \nu\right)$. By definition

$$\lambda(B) = \inf_{f\in c_0(B)} \frac{\left((I - P)^B f, f\right)_\nu}{\|f\|_2^2} \le \frac{\left((I - P)^B v, v\right)_\nu}{\|v\|_2^2}.$$

Let $v(y)$ be a harmonic function on $B\backslash\{a\}$, $v(a) = \rho(a, b)$ and $v(b) = 0$. Again from (3.7)

$$\left((I - P)^B v, v\right) = \rho(a, b),$$

while by using the Cauchy-Schwarz inequality, we have

$$\|v\|_2^2 \geq \frac{E_a(\tau_b)^2}{\nu(\Gamma_a^b)}.$$

∎

Corollary 3.3. *For all random walks on weighted graphs and $R \geq 2, M \geq 2$*

$$\lambda(x, MR) \leq \frac{\rho(x, R, MR) v(x, R, MR)}{E(\underline{w}, \frac{M}{2}R)^2},$$

where $\underline{w} \in \partial B(x, \frac{M}{2}R)$ minimizes $E(w, \frac{M}{2}R)$.

Proof   Let us observe that the walk crosses $S = \partial B\left(x, \frac{MR}{2} - 1\right)$ at $w \in S$ and leaves $B\left(w, \frac{M}{2}R\right)$ before leaving $B\left(x, MR\right)$. (See Figure 3.1.) That means that

$$\min_{w \in S} E(w, \frac{M}{2}R) \leq E_a(\tau_b).$$

∎

For later use, let us introduce the following notation:

$$\underline{E}\left(x, \frac{1}{2}MR\right) = \min_{w \in \partial B(x, \frac{MR}{2} - 1)} E(w, \frac{M}{2}R),$$

which helps to summarize our observations:

$$\underline{E}\left(x, \frac{1}{2}MR\right)^2 \leq \lambda^{-1}(x, MR)\,\rho(x, R, MR)\,v(x, R, MR),$$

and

$$\overline{E}(x, MR) \geq \lambda^{-1}(x, MR).$$

Proposition 3.3. *For all weighted graphs and finite sets $A \subset B \subset \Gamma$*

$$\rho(A, \partial B)\mu(B \backslash A) \geq d(A, \partial B)^2.$$

Proof   The proof follows the idea of Lemma 1 of [94]. Write $L = d(A, \partial B)$, and $S_i = \{z \in B : d(A, z) = i\}, S_0 = A, S_L = \partial B$ and $E_i = \{(x, y) : x \in S_i, y \in S_{i+1}\}, \mu(E_i) = \sum_{(z,w) \in E_i} \mu_{z,w}$

$$\rho(A, \partial B) \geq \sum_{i=0}^{L-1} \rho(S_i, S_{i+1}) = \sum_{i=0}^{L-1} \frac{1}{\mu(E_i)} \geq \frac{L^2}{\sum_{i=0}^{L-1} \mu(E_i)}.$$

∎

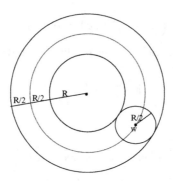

Fig. 3.1. Leaving nested balls

This proposition has an interesting consequence which is in close connection with Lemma 2.2.

Corollary 3.4. *For all weighted graphs and $x \in \Gamma, R > r > 0$,*

$$\rho(x, r, R)v(x, r, R) \geq (R - r)^2, \tag{3.10}$$

$$\rho(x, r, 2r)v(x, r, 2r) \geq r^2. \tag{3.11}$$

Proof   The first statement is a direct consequence of Proposition 3.3, the second follows from the first one.   ∎

Finally, we present a nice result [21] which was little known for some time. Let $\mathcal{C}_{a,b}$ be the commute time between two vertices of a finite graph Γ.

Theorem 3.2. *For any $a, b \in \Gamma$, in a weighted finite graph (Γ, μ)*

$$\mathbb{E}\left(\mathcal{C}_{a,b}\right) = \rho\left(a, b\right)\mu\left(\Gamma\right),$$

where

$$\mathcal{C}_{a,b} = \min\left\{k > T_b : X_k = a | X_0 = a\right\}.$$

The proof is elementary but instructive, so we recall it here.

Proof   Let us use $T_{x,y}$ for the time the walker starting from x needs to reach y. It is clear that $\mathcal{C}_{a,b} = T_{a,b} + T_{b,a}$. Let us inject $\mu(x)$ unit of current in the graph at $x \in \Gamma$ and remove $\mu(\Gamma)$ from b. $v_b(x)$ denote the potential level at x. Using Kirchoff's and Ohm's laws we have

$$\mu\left(x\right) = \sum_{y \sim x}{}'\left(v_b\left(x\right) - v_b\left(y\right)\right)\mu_{x,y} \tag{3.12}$$

for all $x \neq b$. This can be rewritten as

$$\mu(x) = v_b(x)\,\mu(x) - \sum_{y \sim x} v_b(y)\,\mu_{x,y},$$

$$v_b(x) = 1 + \sum_{y \sim x} \frac{\mu_{x,y}}{\mu(x)} v_b(y) = \sum_{y \sim x} \frac{\mu_{x,y}}{\mu(x)}(1 + v_b(y))$$

Meanwhile from the Kolmogorov equation for $H_{x,y} = E\left(T_{x,y}\right)$ we have

$$H_{x,b} = \sum_{y \sim x} \frac{\mu_{x,y}}{\mu(x)}\left(1 + H_{y,b}\right), \tag{3.13}$$

for all $x \neq a$. Since the two systems of non-degenerate linear equations are the same, the solutions are the same as well:

$$H_{x,b} = v_b(x).$$

Reversing the currents, the same argument proves that $H_{a,x} = v_a(x)$. In the superposition of the two systems of currents, the currents cancel each other at all $x \neq a, b$. Consequently, the total voltage between a and b is $H_{a,b} + H_{b,a}$ which is $E\left(\mathcal{C}_{a,b}\right)$ by definition. On the other hand, the total injected current is $\mu\left(\Gamma\right)$, hence by Ohm's law

$$E\left(\mathcal{C}_{a,b}\right) = \rho(a,b)\,\mu\left(\Gamma\right).$$

∎

Lemma 3.1. *On all (Γ,μ) for any $x \in \Gamma, R > S > 0$*

$$E(x, R + S) \geq E(x, R) + \min_{y \in \partial B(x,R)} E(y, S).$$

Proof   Let us write $A = B(x, R)$, $B = B(x, R + S)$. First, let us observe that from the triangular inequality it follows that for any $y \in S(x, R)$

$$B(y, S) \subset B(x, R + S).$$

From this and the strong Markov property we obtain that

$$E(x, R + S) = E_x\left(T_B + E_{X_{T_B}}(x, R + S)\right)$$
$$\geq E(x, R) + E_x\left(E_{X_{T_B}}(X_{T_B}, S)\right).$$

However, $X_{T_B} \in \partial B(x, R)$, from which the statement follows.   ∎

Corollary 3.5. *The mean exit time $E(x, R)$ for $R \in \mathbb{N}$ is strictly monotone and has inverse $e(x, n): \Gamma \times \mathbb{N} \to \mathbb{N}$,*

$$e(x, n) = \min\left\{R \in \mathbb{N} : E(x, R) \geq n\right\}.$$

Proof   Simply let $S = 1$ in Lemma 3.1 and use that $E(z, 1) \geq 1$:

$$E(x, R + 1) \geq E(x, R) + \min_{z \in S(x,R)} E(z, 1) \geq E(x, R) + 1. \tag{3.14}$$

∎

Lemma 3.2. *On all (Γ, μ) for any $x \in A \subset \Gamma$*

$$E_x\left(T_A\right) \leq \rho\left(\{x\}, A^c\right)\mu\left(A\right).$$

Proof   This observation is well known (cf. [94], or [3]), therefore we give the proof in a concise form. It is known that $g^A\left(x, y\right) \leq g^A\left(x, x\right) = \rho\left(\{x\}, A^c\right).$

$$\begin{aligned}
E_x\left(T_A\right) &= \sum_{y \in A} G^A\left(x, y\right) \\
&= \sum_{y \in A} g^A\left(x, y\right)\mu\left(y\right) \\
&\leq g^A\left(x, x\right)\sum_{y \in A}\mu\left(y\right) \\
&= \rho\left(\{x\}, A^c\right)\mu\left(A\right).
\end{aligned}$$

∎

Let $A \subset \Gamma$, Γ^a denote the graph with vertex set $\Gamma^a = A^c \cup \{a\}$, where a is a new vertex added to the vertex set. The edge set contains all edges $x \sim y$ for $x, y \in A^c$ and their weights remain the original $\mu^a_{x,y} = \mu_{x,y}$. There is an edge between $x \in A^c$ and a if there is a $y \in A$ for which $x \sim y$ and the weights are defined by $\mu^a_{x,a} = \sum_{y \in A} \mu_{x,y}$. The random walk on (Γ^a, μ^a) is defined as random walks on weighted graphs are defined in general. The graph Γ^a is obtained by shrinking the set A in a single vertex a.

Corollary 3.6. *For (Γ, μ) and for finite sets $A \subset B \subset \Gamma$ consider (Γ^a, μ^a) and the corresponding random walk.*

$$E_a\left(T_B\right) \leq \rho\left(A, B^c\right)\mu\left(B \backslash A\right). \tag{3.15}$$

Proof   The statement is an immediate consequence of Lemma 3.2.   ∎

Lemma 3.3. *For (Γ, μ) for all $x \in \Gamma, R > 0$,*

$$\min_{z \in \partial B\left(x, \frac{3}{2}R\right)} E\left(z, R/2\right) \leq \rho\left(x, R, 2R\right) v\left(x, R, 2R\right). \tag{3.16}$$

Proof   Let us consider the annulus $D = B\left(x, 2R\right) \backslash B\left(x, R\right)$. Let us apply Corollary 3.6 to $A = B\left(x, R\right), B = B\left(x, 2R\right)$ to receive

$$\rho\left(x, R, 2R\right) v\left(x, R, 2R\right) \geq E_a\left(T_B\right). \tag{3.17}$$

It is clear that the walk starting in a and leaving B should cross $\partial B\left(x, 3/2R\right)$. Now we use the Markov property as in Lemma 3.1. Denote the first hitting (random) point by ξ. It is evident that the walk continuing from ξ should leave $B\left(\xi, \frac{1}{2}R\right)$ before it leaves $B\left(x, 2R\right)$. That means that

$$\rho\left(x, R, 2R\right) v\left(x, R, 2R\right)$$

$$\geq E_a\left(T_B\right) \geq \min_{y \in \partial B\left(x, \frac{3}{2}R\right)} E\left(y, \frac{1}{2}R\right).$$

∎

Corollary 3.7. *If $(p_0),(VD)$ and $(\overline{E})$ hold, then there is a $c > 0$ such that for all $x \in \Gamma, R > 0$*

$$E(x, R) \geq cR^2, \tag{3.18}$$

and if for all $x \in \Gamma, R > 0$ and fixed $\beta, C > 0$

$$E(x, R) \leq CR^\beta,$$

then

$$\beta \geq 2.$$

Proof   The statement follows easily from Lemmas 2.3 and 2.2.   ∎

Remark 3.1. Since, it is clear that $(TC) \Longrightarrow (\overline{E})$ we also have the implication $(p_0),(VD),(TC) \Longrightarrow (3.18)$.

Exercise 3.3. The following statements are equivalent

1. There are $C, c > 0, \beta \geq \beta' > 0$ such that for all $x \in \Gamma, R \geq r > 0$, $y \in B(x, R)$

$$c\left(\frac{R}{r}\right)^{\beta'} \leq \frac{E(x, R)}{E(y, r)} \leq C\left(\frac{R}{r}\right)^\beta, \tag{3.19}$$

2. There are $C, c > 0, \beta \geq \beta' > 0$ such that for all $x \in \Gamma, n \geq m > 0$, $y \in B(x, e(x, n))$

$$c\left(\frac{n}{m}\right)^{1/\beta} \leq \frac{e(x, n)}{e(y, m)} \leq C\left(\frac{n}{m}\right)^{1/\beta'}. \tag{3.20}$$

3.3 Harnack inequality and the Green kernel

Definition 3.2. *We say that a weighted graph (Γ, μ) satisfies the elliptic Harnack inequality $(\mathbf{H})$ if, for all $x \in \Gamma, R > 0$ and for any non-negative function u which is harmonic in $B(x, 2R)$, the following inequality holds*

$$\max_{B(x,R)} u \leq H \min_{B(x,R)} u, \tag{3.21}$$

with some constant $H > 1$ independent of x and R.

Definition 3.3. *We say that a weighted graph (Γ, μ) satisfies the elliptic Harnack inequality with shrinking parameter $M > 1$ and we refer to it as $H(M)$, if for all $x \in \Gamma, R > 0$ and for any non-negative harmonic function u which is harmonic in $B(x, MR)$, the following inequality holds*

$$\max_{B(x,R)} u \leq H \min_{B(x,R)} u \tag{3.22}$$

with some constant $H > 1$ independent of x and R. For convenience, the condition $H(2)$ will be denoted by (H).

Remark 3.2. It is easy to see that for any fixed R_0 for all $R < R_0$ the Harnack inequality follows from (p_0).

In this section we establish a connection between the elliptic Harnack inequality and the regular behavior of Green functions.

We consider the following Harnack inequality for Green functions.

Definition 3.4. *We say that* (Γ, μ) *satisfies* **wHG** (U, M) *the weak Harnack inequality for Green functions if there are constants* $M \geq 2, C \geq 1$ *such that for all* $x \in \Gamma$ *and* $R > 0$ *and for any finite set* $U \supset B(x, MR)$,

$$\sup_{y \notin B(x,R)} g^U(x, y) \leq C \inf_{z \in B(x,R)} g^U(x, z). \tag{3.23}$$

Definition 3.5. *We say that* (Γ, μ) *satisfies* **HG** (U, M) *the Harnack inequality for Green functions if there are constants* $M \geq 2, C \geq 1$ *such that for all* $x \in \Gamma$ *and* $R > 0$ *and for any finite set* $U \supset B(x, MR)$,

$$\sup_{y \notin B(x,R/2)} g^U(x, y) \leq C \inf_{z \in B(x,R) \setminus B(x,R/2)} g^U(x, z). \tag{3.24}$$

Definition 3.6. *We say that* (Γ, μ) *satisfies* **HG** (M) *(or simply* **HG***) the annulus Harnack inequality for Green functions for balls if* $U = B(x, 2R)$ *is a ball for* $x \in \Gamma, R > 0$.

Remark 3.3. From the maximum principle it follows that we receive a definition equivalent to **HG** (U, M) if we replace (3.24) by

$$\sup_{y \in B(x,R) \setminus B(x,R/2)} g^U(x, y) \leq C \inf_{z \in B(x,R) \setminus B(x,R/2)} g^U(x, z). \tag{3.25}$$

Proposition 3.4. *Assume that* (p_0) *holds on* (Γ, μ). *Then*

$$HG(M) \implies H(6M).$$

The main part of the proof is contained in the following lemma.

Lemma 3.4. *Let* $B_0 \subset B_1 \subset B_2 \subset B_3$ *be a sequence of finite sets in* Γ *such that* $\overline{B_i} \subset B_{i+1}$, $i = 0, 1, 2$. *Let* $A = \overline{B_2} \setminus B_1$, $B = B_0$ *and* $U = B_3$. *Then for any non-negative harmonic function* u *in* B_2,

$$\max_B u \leq H \inf_B u, \tag{3.26}$$

where

$$H = \max_{x \in B} \max_{y \in B} \max_{z \in A} \frac{G^U(y, z)}{G^U(x, z)}. \tag{3.27}$$

Remark 3.4. Note that no assumption is made about the graph. If the graph is transient, that is, the global Green kernel $G(x, y)$ is finite, then by exhausting Γ by a sequence of finite sets U, we can replace G^U in (3.27) by G.

Proof   The following potential-theoretic argument was borrowed from [17]. Given a non-negative harmonic function u in B_2, let S_u denote the class of all non-negative functions v on $\overline{U}$ such that v is super-harmonic in U and $v \geq u$ in $\overline{B_1}$. Define the function w on $\overline{U}$ by

$$w(x) = \min \{v(x) : v \in S_u\}.$$

Clearly, $w \in S_u$. Since the function u itself is in S_u too, we have $w \leq u$ in $\overline{U}$. On the other hand, $w \geq u$ in $\overline{B_1}$, whence we see that $u = w$ in $\overline{B_1}$. In particular, it suffices to prove (3.26) for w instead of u.

Let us show that $w \in c_0(U)$, i.e., $w(x) = 0$ if $x \in \overline{U} \setminus U$. For any given $x \in \overline{U} \setminus U$, let us construct a barrier function $v \in S_u$, such that $v(x) = 0$. Such a function v can be obtained as the solution of a Dirichlet boundary value problem in U, with any positive data on $\overline{U} \setminus U$, except for point x, where the boundary function should vanish. By the strong minimum principle, $v > 0$ in U. Therefore, for a large enough constant C, we have $Cv \geq u$ in $\overline{B_1}$ which implies $Cv \geq w$ in $\overline{U}$ and, consequently, $w(x) = 0$.

Write $f = -\Delta w \geq 0$. Since $w \in c_0(U)$, for any $x \in U$ we have

$$w(x) = \sum_{z \in U} G^U(x, z) f(z). \tag{3.28}$$

We claim that w is harmonic outside of A, that is, $f = 0$ in $U \setminus A$. If $x \in B_1$, then

$$f(x) = -\Delta w(x) = -\Delta u(x) = 0$$

because $w = u$ in $\overline{B_1}$. Let $x \in U \setminus \overline{B_2}$ and assume that $f(x) > 0$, i.e.,

$$w(x) > Pw(x) = \sum_{y \sim x} P(x, y) w(y).$$

Consider the function $\overline{w}$ which is equal to w everywhere in $\overline{U}$ except for point x, and define $\overline{w}$ at x as

$$\overline{w}(x) = \sum_{y \sim x} P(x, y) w(y).$$

Clearly, $\overline{w} \in S_u$. Hence, by the definition of w, we have $w \leq \overline{w}$ which contradicts $w(x) > \overline{w}(x)$.

Since $f = 0$ in $U \setminus A$, the summation in (3.28) can be restricted to $z \in A$, whence, for all $x, y \in B$,

$$\frac{w(y)}{w(x)} = \frac{\sum_{z \in A} G^U(y, z) f(z)}{\sum_{z \in A} G^U(x, z) f(z)} \leq H,$$

and (3.26) follows.   ∎

Proof   [of Proposition 3.4] Now we assume $HG(M)$ and prove $H(6M)$. Let $p \in \Gamma, R > 1$ and $U = B(p, 6MR), B = B(p, R), A = B(p, 5R) \setminus B(p, 4R)$ and $u \geq 0$ be a harmonic function on U. From Lemma 3.4 we have that

$$\max_B u \leq H \min_B u,$$

where

$$H = \max_{x,y \in B} \max_{z \in A} \frac{g^U(x, z)}{g^U(y, z)}.$$

We show that $H < C_{HG}$. If $x, y \in B(p, R)$ and $z \in A$, we can see that $x \in B(z, 3R)^c$ and $y \in B(z, 6R)$, furthermore $5R + 3MR < 6MR$ and $B(z, 3MR) \subset U$, and we can apply $HG(M)$ to obtain

$$\frac{g^U(x, z)}{g^U(y, z)} \leq C_{HG}.$$

This shows $H < C_{HG}$.   ∎

Proposition 3.5. *Assume that (Γ, μ) satisfies (p_0). Then for $M \geq N$,*

$$H(N) \Longleftrightarrow H(M).$$

Proof   It is clear that $H(N) \implies H(M)$ since if $u \geq 0$ is harmonic in $B(x, MR)$, then by $H(N)$

$$\max_{B(x,R)} u \leq \max_{B(x, \frac{M}{N}R)} u \leq C_N \min_{B(x, \frac{M}{N}R)} u \leq C_N \min_{B(x,R)} u.$$

We show the opposite direction $H(M) \implies H(N)$. Assume that $u \geq 0$ is harmonic on $B(x, MR)$, write $B = B(x, R)$. We have by assumption that

$$u(x_1) = \max_B u \leq C_M \min_B u = u(x_2). \tag{3.29}$$

and writing $D = \partial B\left(x, \frac{M}{N}R\right)$ we obtain that

$$\max_D u = u(y_1),$$

$$\min_D u = u(y_2).$$

Let $z_i \in \partial B(x, R)$ be the intersection of the shortest path π_i from x to y_i. It is then clear from (3.29) that

$$u(z_1) \leq C_M u(z_2). \tag{3.30}$$

Let us consider the minimal chain of intersecting balls between z_1 and y_1 in the following way. $B_j = B(o_j, r), o_j \in \pi_1, r = \frac{M\left(1 - \frac{1}{N}\right) - 1}{M - 1} R$ such that $z_1 \in B_1$

and $y_1 \in B_K$. It is clear that $K \leq \frac{(\frac{M}{N}-1)R}{2(r-1)} \leq 2M$. The choice of r ensures that $B(o_j, Mr) \subset B(x, MR)$ and $H(M)$ can be applied to compare the values $u(z_1) \geq \frac{1}{C_M} u(o_1) \ldots \geq C_M^{-K-2} u(y_1)$. Similarly $u(z_2) \leq C_M^{K+2} u(y_2)$. Substituting these inequalities in (3.30) we obtain that

$$u(y_1) \leq (C_M)^{4M+5} u(y_2).$$

∎

Corollary 3.8. *Assume that (Γ, μ) satisfies (p_0). Then for any $M \geq 2$*

$$HG(M) \Longrightarrow (H).$$

Proof   We have seen in Proposition 3.4 that $HG(M) \Longrightarrow H(6M)$ and in Proposition 3.5 that $H(6M) \Longrightarrow H(2)$.   ∎

In the sequel the converse statement is needed.

Lemma 3.5. *Assume that (Γ, μ) satisfies (p_0). Then for any $M \geq 2$*

$$(H) + wHG(U, 2) \Longrightarrow HG(U, M).$$

Exercise 3.4. The proof is left to the reader as an exercise. Use the chaining argument along rays as above.

Proposition 3.6. *Assume that (Γ, μ) satisfies (p_0). Then in the case when $M \geq 14$*

$$(BC) + (H) \Longrightarrow wHG(U, M).$$

In particular,

$$(VD) + (H) \Longrightarrow wHG(U, M).$$

Lemma 3.6. *Assume that (Γ, μ) satisfies (p_0), (BC), (H). Let y, z be two points in a ball $B(x, R)$, such that the shortest path*

$$y = \xi_0 \sim \xi_1 \sim \ldots \sim \xi_k = z \tag{3.31}$$

connecting y and z in Γ does not intersect $B(x, \varepsilon R)$ for some $\varepsilon \in (0, 1)$. Then for any finite set U containing $B(x, 3R)$,

$$g^U(x, y) \leq C_\varepsilon g^U(x, z). \tag{3.32}$$

Proof   If R is in a bounded range, then (3.32) follows from the hypothesis (p_0) (cf. Remark 3.2). Assume in the sequel that R is large enough, and observe that for any ξ_i

$$d(x, \xi_i) \leq \frac{d(x, y) + d(x, z) + d(y, z)}{2} \leq d(x, y) + d(x, z) \leq 2R,$$

so that $\xi_i \in B(x, 2R)$. By (BC), the ball $B(x, 2R)$ can be covered at most by $N(\varepsilon/20)$ number of balls of radius $r = \frac{\varepsilon}{10}R$. Select only the balls which contain at least one ξ_i, and denote their centers by o_j, $j = 0, 1, ..., n$, where $n \leq N(\varepsilon/20)$. Clearly, it is possible to enumerate the selected balls so that $B(o_0, r)$ contains y, $B(o_n, r)$ contains z, and $d(o_j, o_{j+1}) \leq 2r+1$ for all $j < n$. Each ball $B(o_j, r)$ contains at least one of the points ξ_i; select one of them and denote it by $\overline{\xi}_j$. Since $d(x, \overline{\xi}_j) \geq \varepsilon R = 10r$, the point x is outside the ball $B(\overline{\xi}_j, 10r)$. Since $10r < R$ and hence $B(\overline{\xi}_j, 10r) \subset B(x, 3R) \subset U$, the function $g^U(x, \cdot)$ is harmonic in $B(\overline{\xi}_j, 10r)$. Since $\overline{\xi}_{j+1} \in B(\overline{\xi}_j, 5r)$, the Harnack inequality (H) yields $g^U(x, \overline{\xi}_j) \leq Cg^U(x, \overline{\xi}_{j+1})$, whence (3.32) follows by iteration. ∎

Proof [of Proposition 3.6] Let y and z be the points where the sup and inf in (HG) are attained, respectively. It follows from the maximum principle, that $z \in B(x, R) \setminus B(x, R-1)$ and $y \in B(x, R+1) \setminus B(x, R)$. We need to prove that

$$g^U(x, y) \leq Cg^U(x, z), \qquad (3.33)$$

where $U \supset B(x, MR)$ for some M. In fact, any $M \geq 14$ will do, as we shall see below. For a bounded range of R, (3.33) follows from (p_0) (cf. Remark 3.2). Assume in the sequel that R is large enough, and connect y and z by the shortest path in Γ as in (3.31).

CASE 1. All points ξ_i are outside $B(x, \frac{1}{4}R)$. Then (3.33) follows from Lemma 3.6 since $y, z \in B(x, 2R)$ and $U \supset B(x, 6R)$.

CASE 2. One of the points ξ_i is in the ball $B(x, \frac{1}{4}R)$. Let us show that the shortest path connecting x and z in Γ does not intersect $B(y, \frac{1}{4}R)$. Indeed, let ξ denote the point ξ_i, such that $d(x, \xi) < \frac{1}{4}R$. By triangle inequality, we have

$$d(y, z) = d(y, \xi) + d(z, \xi) \geq (d(x, y) - d(x, \xi)) + (d(x, z) - d(x, \xi)) \quad (3.34)$$

$$> d(x, y) + d(x, z) - \frac{R}{2}. \qquad (3.35)$$

For any point η on the shortest path between x and z, by (3.34) we obtain

$$d(y, \eta) \geq d(y, z) - d(\eta, z) \geq d(y, z) - d(x, z) \geq d(x, y) - \frac{R}{2} > \frac{R}{4}.$$

Hence, the path from x to z is outside the ball $B(y, \frac{1}{4}R)$. Since $x, z \in B(y, 4R)$ and $U \supset B(y, 12R)$ by Lemma 3.6 we obtain $g^U(y, x) \leq Cg^U(y, z)$. Similarly, connecting x to y, we obtain $g^U(z, y) \leq Cg^U(z, x)$. Multiplying the inequalities we obtain (3.33). Finally we know that $(VD) \Longrightarrow (BC)$. ∎

Corollary 3.9. *Assume that (Γ, μ) satisfies (p_0). Then for $M \geq 2$*

$$(BC) + (H) \Longrightarrow HG(U, M)$$

In particular,

$$(VD) + (H) \Longrightarrow HG(U, M).$$

Proof   The statement follows from Proposition 3.6 and Lemma 3.5. ∎

Remark 3.5. The statement of Proposition 3.6 is improved by Barlow in [4], as it is given in the following Lemma.

Lemma 3.7. *Assume that* (Γ, μ) *satisfies* (p_0), *then*

$$H(2) \implies wHG(U, 2). \tag{3.36}$$

Proof   Let us assume that $R > 12$ and is divisible by 12. Denote y, z the points where the maximum and minimum attained. Let $y\prime$ denote the midpoint of the shortest path from x to y and z' the midpoint from x to z. That means that $d(x, y') = d(x, z') = R/2$ and $d(y', z)$ and $d(z', y) \geq R/2$. We separate two cases.

1. Assume that $d(y', z') \leq R/3$. Let v be as close as possible to the midpoint of the shortest path between $y\prime$ and $z\prime$. We have that $d(v, y') \leq 1 + R/6 \leq R/4$ and (H) can be applied to $g^B(x, .)$ in $B(y', R/4)$. We obtain that

$$cg^U(x, y') \leq g^U(z, v) \leq Cg^U(x, y').$$

   Now we apply (H) to $g^U(x, .)$ in $B(y, R/2)$:

$$cg^U(x, y) \leq g^U(x, y') \leq g^U(x, y).$$

   The result of the combination of these inequalities is that

$$cg^U(x, y) \leq g^U(x, v) \leq g^U(x, y),$$

   and the same arguments applied to $g^U(x, z)$ prove the statement.
2. Assume that $d(y', z') > R/3$. In this case apply (H) to $g^U(z, .)$ in $B(x, R/2)$:

$$cg^U(z, y') \leq g^U(z, x) \leq Cg^U(z, y'). \tag{3.37}$$

   Let $v\prime$ be on the shortest path between z', z with $d(z', v') = s, 0 \leq s \leq R/2$. Since $d(y', z') > R/3$ and $d(y', z) \geq R/2$, we have that $d(y'v') \geq \max\{R/3 - s, s\}$. This means that $d(y', v') \geq R/6$. We can apply (H) to $g^U(y', .)$ along a chain of balls to deduce

$$cg^U(y', z') \leq g^U(y', z) \leq Cg^U(y', z'). \tag{3.38}$$

   From the relations (3.37) and (3.38) it follows that

$$g^U(z, x) \leq C_1 g^U(z, y') \leq C_2 g^U(y', z'),$$
$$g^U(y', z') \leq C_3 g^U(z, y') \leq C' g^U(z, x).$$

   The same can be deduced for $g^U(y, x)$ and $g^U(y', z')$ which, using the maximum (minimum principle), leads to (wHG). If $R < 12$, the statement follows from (p_0). If R is not divisible by 12 let $R' = 12k, R - 11 \leq R' < R$ and find points $y'', z'' \in \partial B(x, R)$ so that $d(y, y'') \leq 11$ and $d(z, z'') \leq 11$. Let us use the above argument for y'' and z'', then by (p_0) we receive the statement. ■

Because of Lemma 3.5 the following corollary holds.

Corollary 3.10. *Assume that* (Γ, μ) *satisfies* (p_0). *Then for* $M \geq 2$

$$(H) \Longrightarrow HG(U, M).$$

Theorem 3.3. *Assume that* (Γ, μ) *satisfies* (p_0). *Then for* $M \geq 2, U \subset \Gamma$

$$(H) \Longleftrightarrow HG(U, M).$$

Proof   We have from Corollary 3.10 that $(H) \Longrightarrow HG(U, M)$ and the reverse implication follows from Corollary 3.8. ∎

Proposition 3.7. *Assume that the graph* (Γ, μ) *satisfies* (p_0) *and* $wHG(2)$. *Then for any ball* $B(x, R)$ *and for any* $0 < r \leq R/2$, *we have*

$$\sup_{y \notin B(x,r)} g^{B(x,R)}(x, y) \simeq \rho(B(x,r), B(x,R)) \simeq \inf_{y \in B(x,r)} g^{B(x,R)}(x, y). \quad (3.39)$$

Proof   For an arbitrary graph (Γ, μ), the following is true: if A, B are finite subsets of Γ such that $\overline{A} \subset B$, then for any $x \in A^\circ$

$$\sup_{y \notin A^\circ} g^B(x, y) \geq \rho(A, B) \geq \inf_{y \in \overline{A}} g^B(x, y), \quad (3.40)$$

where A° consists of the points in A having neighbors only in A (see Proposition 3.1). Applying (3.40) for $A = B(x, r)$, $B = B(x, R)$ and combining it with $wHG(2)$ and (p_0), we obtain (3.39). ∎

Proposition 3.8. *Assume that the graph* (Γ, μ) *satisfies* (p_0) *and* $wHG(2)$. *Fix any ball* $B(x, r)$ *and write* $B_k = B(x, 2^k r)$ *for* $k = 0, 1, \dots$. *Then for all integers* $n > m \geq 0$,

$$\sup_{y \notin B_m} g^{B_n}(x, y) \simeq \sum_{k=m}^{n-1} \rho(B_k, B_{k+1}) \simeq \inf_{y \in B_m} g^{B_n}(x, y). \quad (3.41)$$

Proof   From the monotonicity principle, in particular from 4 of Corollary 3.2, it follows that shrinking the boundary of B_is we get

$$\sum_{k=m}^{n-1} \rho(B_k, B_{k+1}) \leq \rho(B_m, B_n).$$

Together with Proposition 3.7, it implies the lower bound for $\inf g^{B_n}$ in (3.41). To obtain the upper bound for $\sup g^{B_n}$, observe that the difference

$$g^{B_{k+1}}(x, \cdot) - g^{B_k}(x, \cdot)$$

is a harmonic function in B_k. The maximum principle implies for any $y \in \Gamma$ that

$$g^{B_{k+1}}(x,y) - g^{B_k}(x,y) \le \sup_{z \notin B_k} g^{B_{k+1}}(x,z).$$

By Proposition 3.7, we obtain

$$g^{B_{k+1}}(x,y) - g^{B_k}(x,y) \le C\rho\left(B_k, B_{k+1}\right). \tag{3.42}$$

For any $y \notin B_m$, Proposition 3.7 yields

$$g^{B_{m+1}}(x,y) \le C\rho(B_m, B_{m+1}). \tag{3.43}$$

From (3.43) and (3.42) for $m < k < n$ we obtain the upper bound of $\sup g^{B_n}$ in (3.41). $\blacksquare$

Remark 3.6. The main observation on the Harnack inequality and the Green kernel is the following. The Harnack inequality implies that the resistance of annuli have the doubling property. The relation (3.39) shows that the elliptic Harnack inequality implies some kind of spherical symmetry of harmonic functions.

3.4 Resistance regularity

Theorem 3.4. *Assume that (Γ, μ) satisfies $(p_0), (H)$. Then*

$$\rho(x, R, 4R) \le C_2 \rho(x, 2R, 4R), \tag{3.44}$$

and, if in addition (BC) holds, then

$$\rho(x, R, 4R) \le C_1 \rho(x, R, 2R) \tag{3.45}$$

where $C_i > 1$ are independent of $x \in \Gamma$ and $R \ge 0$.

Proof   Assume that (Γ, μ) satisfies $(p_0), (H)$. If $R \le 16$, the statement follows from (p_0), so we assume that $R > 16$. The first statement is a direct consequence of (H) and (3.39). Since Γ is connected there is a path from x to $B^c(x, 4R)$. This path has an intersection with $\partial B(x, R)$ in y_0 and with $\partial B(x, 2R - 2)$ in z_0. Along this path we can form a finite intersecting chain of balls $B(x_i, R/4)$ with centers on the path starting with $x_0 = y_0$ and ending with $x_K = z_0$. It is clear that $x \notin B(x_i, R/2) \subset B(x, 4R) =: B$ hence $g^B(x,.)$ is harmonic in them and the Harnack inequality and the standard chaining argument can be used to obtain

$$g^B(x, y_0) \le Cg^B(x, z_0).$$

Now using (3.39) twice, it follows that

$$\rho\left(x, R, 4R\right) \leq C \inf_{y \in B(x,R)} g^{B}\left(x, y\right)$$

$$\leq Cg^{B}\left(x, y_{0}\right) \leq Cg^{B}\left(x, z_{0}\right)$$

$$\leq C \sup_{z \notin B(x,2R)} g^{B}\left(x, z\right)$$

$$\leq C\rho\left(x, 2R, 4R\right).$$

Let us prove (3.45). Let $U = B\left(x, 5R\right)$, $A = B\left(x, R\right)$, $D = B\left(x, 4R\right) \setminus B\left(x, \frac{3}{2}R\right)$. Consider the connected components of D, denote them by D_{i} and $S_{i}\left(x, r\right) = S\left(x, r\right) \cap D_{i}$. From the bounded covering condition it follows that the number of these components is bounded by K. Let $\Gamma_{i} = D_{i} \cup \left[B\left(x, 5R\right) \setminus B\left(x, 4R\right)\right] \cup B\left(x, \frac{3}{2}R\right)$ (see Figure 3.2).

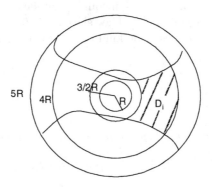

Fig. 3.2. Connected components in the annulus

It is clear that

$$\frac{1}{\rho(x, R, 2R)} \leq \sum_{i=1}^{K} \frac{1}{\rho(B\left(x, R\right), S_{i}\left(x, 2R\right))}$$

$$\leq \frac{K}{\min_{i} \rho(B\left(x, R\right), S_{i}\left(x, 2R\right))}.$$

Let us assume that the minimum is obtained for $i = 1$, so that

$$\frac{\rho(x, R, 4R)}{\rho(x, R, 2R)} \leq K \frac{\rho(B\left(x, R\right), S_{1}\left(x, 4R\right))}{\rho(B\left(x, R\right), S_{1}\left(x, 2R\right))}. \tag{3.46}$$

Let us consider the capacity potential $u\left(y\right)$ between $B^{c}\left(x, 5R\right)$ and $B\left(x, R\right)$ which is set zero on $B\left(x, R\right)$ and $u\left(w\right) = \rho(x, R, 5R)$ for $w \in B^{c}\left(x, 5R\right)$. It is clear that $u\left(y\right)$ is harmonic in D. Then our strategy is the following. We will compare potential values of u by using the Harnack inequality along a chain of balls again consisting of a bounded number of balls. Thanks to the bounded covering property, $B(x, 5R)$ can be covered by a bounded number of balls of radius $r = R/16$. We consider the subset of those balls which are intersecting D_1. If $B_i = B(o_i, r)$ is such a ball, it is clear that $B(o_i, 4r)$ does not intersect $B(x, R)$ and $B^c(x, 5R)$, and hence u is harmonic in $B(o_i, 4r)$. First, let $y, y' \in D_1$ and

$$\pi = \pi(y, y') = \{y_0, y_1, \dots y_N = y'\},$$

the shortest path connecting them.

Let us consider a minimal covering of the path by balls. Let us pick up the ball $B(o_i, r)$ of the smallest index which contains y, then fix the last point along the path from y to $y_1 \in \pi(y, y')$ which is within this ball and the next one, z_1, which is not. Now let us pick up the ball with the smallest index $B(o_j, r)$ which covers z_1. From the triangular inequality

$$d(o_i, z_1) \leq d(z_1, y_1) + 1 = r + 1,$$

it follows that $z_1 \in B(o_i, 2r)$ if $r \geq 1$, which means that the elliptic Harnack inequality applied in $B(o_i, 4r)$ and then in $B(o_j, 4r)$, implies that

$$u(y) \leq Cu(z_1).$$

The procedure can be continued until either y' is covered or all balls are used. Since at least one new point is covered in each step and only unused balls are selected, the procedure has no loop in it. We use at most K balls to cover $B(x, 5R)$. Of course, when y' is covered we are ready since at most K balls were used and $K + 1$ iterations must be made. It means that

$$u(y) \leq C^{K+1} u(y'). \tag{3.47}$$

Let us apply this comparison to $y, y \prime \in S_1(x, 2R)$, then to $z, z \prime \in S_1(x, 4R)$. From the maximum principle it follows that

$$\min_{y \in S(x, 2R)} u(y) \leq \rho(B(x, R), S_1(x, 2R)) \leq \max_{y \in S(x, 2R)} u(y).$$

This, together with (3.47), results in

$$\rho(B(x, R), S_1(x, 2R)) \simeq u(y)$$

for all $y \in S_1(x, 2R)$. The same argument yields that

$$\rho(B(x, R), S_1(x, 4R)) \simeq u(z)$$

for all $z \in S_1(x, 4R)$. Finally let us consider a ray from x to a $z_0 \in S_1(x, 4R)$ and its intersection y_0 with $S_1(x, 2R)$. This ray gives the shortest path between y_0 and z_0, and another chaining gives that

$$\rho(B(x, R), S_1(x, 4R)) \simeq u(z_0) \simeq u(y_0) \simeq \rho(B(x, R), S_1(x, 2R))$$

which, by (3.46), gives the statement.   ■

Remark 3.7. In the rest of this section the constants C_1, C_2 refer to the fixed constants of (3.44) and (3.45).

Corollary 3.11. *Under the conditions of Theorem 3.4 the inequalities*

$$\rho(x, 2R, 4R) \leq (C_1 - 1)\, \rho(x, R, 2R), \tag{3.48}$$

$$\rho(x, R, 2R) \leq (C_2 - 1)\, \rho(x, 2R, 4R), \tag{3.49}$$

and

$$\frac{1}{C_1 - 1}\rho(x, 2R, 4R) \leq \rho(x, R, 2R) \leq (C_2 - 1)\, \rho(x, 2R, 4R) \tag{3.50}$$

hold.

Proof   The first two statements follow in the same way from the easy observation that

$$\rho(x, R, 4R) \geq \rho(x, R, 2R) + \rho(x, 2R, 4R).$$

∎

We end this section with some observations which might be interesting on their own. Some consequences of (3.48) and (3.49) are derived (in fact consequences of the elliptic Harnack inequality) and bounds on the volume growth and the mean exit time are deduced.

Remark 3.8. Let us observe that from (3.48) it follows that the graph is transient if $C_1 < 2$, and in this case (3.49) ensures that decay is not faster than polynomial. Similarly from (3.49) it follows that the graph is recurrent if $C_2 \leq 2$, and (3.48) ensures that the resistance does not increase faster than polynomial. It is clear from (3.50) that

$$(C_1 - 1)(C_2 - 1) \geq 1.$$

Remark 3.9. From the (p_0), (3.48) and (3.49) we can easily obtain the following inequalities.

$$\rho(x, R, 2R) \leq C\rho(x, 1, 2)\, R^{\log_2(C_1 - 1)} \leq \frac{C}{\mu(x)} R^{\log_2(C_1 - 1)}, \tag{3.51}$$

$$\rho(x, R, 2R) \geq C\rho(x, 1, 2)\, R^{-\log_2(C_2 - 1)} \geq \frac{c}{\mu(x)} R^{-\log_2(C_2 - 1)}. \tag{3.52}$$

Remark 3.10. Barlow in [4] proved that (p_0) and the elliptic Harnack inequality imply

$$|V(x, R)| \leq CR^{1+\theta}, \tag{3.53}$$

where $\theta = \log_3 H$ and H is the constant in the Harnack inequality. The combination of Corollary 3.4 and (3.51) implies a lower bound for volume growth:

$$R^2 \leq (V(x, 2R) - V(x, R)) \rho(x, R, 2R) \leq V(x, 2R) \frac{C}{\mu(x)} R^{\log_2(C_1 - 1)},$$

therefore, $(p_0), (H)$ and (BC) imply a lower bound for V:

$$V(x, 2R) - V(x, R) \geq c\mu(x) R^{2 - \log_2(C_1 - 1)}. \tag{3.54}$$

In particular, if $V(x, R) \leq CR^\alpha$, then

$$C_1 \geq 2^{2 - \alpha} + 1. \tag{3.55}$$

Similarly, by (H) and (3.52)

$$E(x, 2R) \geq c\rho(x, R, 2R) V(x, R) \geq cV(x, R) R^{-\log_2(C_2 - 1)}, \tag{3.56}$$

which means that $(p_0) + (H)$ implies an upper bound for V:

$$V(x, R) \leq CE(x, R) \mu(x) R^{\log_2(C_2 - 1)}.$$

Similarly to (3.55), we get

$$C_2 \geq 2^{\alpha_1 - \beta} + 1, \tag{3.57}$$

if we assume that $E(x, R) \leq CR^\beta$ and $V(x, R) \geq c\mu(x) R^{\alpha_1}$.

Remark 3.11. We can restate the above observations starting from (3.11) and using (3.49) and (H).

$$\begin{aligned} R^2 &\leq \rho(x, R, 2R) (V(x, 2R) - V(x, R)) \\ &\leq (C_2 - 1) \rho(x, 2R, 4R) V(x, 2R) \\ &\leq (C_2 - 1) E(x, 4R). \end{aligned}$$

This means that $(p_0) + (H)$ implies

$$E(x, R) \geq cR^2.$$

The following corollary highlights the connection between volume growth and resistance properties implied by the elliptic Harnack inequality. In particular, an upper bound for the volume of a ball (similar to the one given in [4]) is provided and complemented with a lower bound.

Corollary 3.12. *Assume that (Γ, μ) satisfies (p_0) and (H). Then there are constants $C, c > 0, C_1, C_2 > 1$ and $\gamma_2 = \log_2(C_2 - 1)$ such that for all $x \in \Gamma, R > 0$*

$$V(x, R) \leq CE(x, R) \mu(x) R^{\gamma_2}$$

and

$$E(x, R) \geq cR^2.$$

In addition if (BC) is satisfied, then there is a $\gamma_1 = \log_2(C_1 - 1)$ such that

$$V(x, 2R) \geq c\mu(x) R^{2 - \gamma_1}.$$

4

Isoperimetric inequalities

Isoperimetry is one of the oldest, if not the oldest, variational problem of mathematics. Zenodorus (who lived not much later than Archimedes) is credited by Pappus to be the first mathematician to study the classical isoperimetric problem.

> *Out of simple closed planar curves with fixed length it is the circle that encloses the largest area.*

This planar problem and its spatial analogue was already studied by Zenodorus with some success. It took almost two thousand years to find rigorous proof. It was done by Schwartz (1884) after the first steps taken by Steiner in 1836.

The importance of isoperimetric inequalities in physics was realized by Lord Rayleigh as early as 1877. He conjectured that the principal frequency of a membrane of fixed area is minimal for the circle. Several further quantities joined this family, including moment of inertia, electric capacity and torsion rigidity. For the history and modern foundation of the topic the reader is advised to study [80],[79] or [21]. It turned out that isoperimetric inequalities are in close connection with ultra-contractive semigroups, in particular with diagonal upper estimates of the heat kernel ([102], [103], [19], [41]). All these works demonstrate that functional analytic inequalities (Sobolev, Nash or Poincaré inequality) and isoperimetric inequalities are closely related. Let us mention here only a few classical results to indicate the role and importance of isoperimetric inequalities in obtaining diagonal upper estimates.

Theorem 4.1. *On a transient graph (Γ, μ), the following statements are equivalent*

1.

$$p_n(x,x) \leq Cn^{-p/2};$$

2. for $f \in c_0(\Gamma), f \geq 0$ the Sobolev inequality $(S_{p/2})$ holds

$$\left[\sum_x f(x)^q \mu(x)\right]^{1/q} \le C \sum_{x,y} \mu_{x,y} \left(f(x) - f(y)\right)^2 ; \qquad (4.1)$$

where $q = \frac{p}{p-2}, p > 2$;
3. for all $A \subset \Gamma$ the Faber-Krahn inequality $\left(FK_{p/2}\right)$

$$\lambda(A)^{-1} \le C\mu(A)^{-\frac{2}{p}}$$

holds.

The equivalence of 1. and 2. is due to Varopoulos [101], the equivalence of 1. and 3. is given by Carron [19] and Grigor'yan [40]. For further results and discussions see [28], [10], [41] or [20].

4.1 An isoperimetric problem

We consider the following isoperimetric problem.

Let (Γ, μ, d) be a measure metric space and assume that diffusion is defined by a strictly local, regular Dirichlet form on it (cf. [36]). Let us fix a starting point $x \in \Gamma$ and $E_x(T_A)$ for a set $A \subset \Gamma$. The following question can be raised.

Which one of *the sets of fixed volume maximizes the mean exit time of the process?*

We establish a partial answer to this question in the context of transient random walks on weighted graphs.

By defining an isoperimetric (or Faber-Krahn) function, the problem is inserted in the set of equivalent conditions of the diagonal upper estimate on a weighted graph. Similar isoperimetric results on the mean exit time are given in [76] under strong conditions and very similar results are given by Carron [19] for continuous space and by Mathieu [74],[75] for continuous time. More specific references to their results will be given after the corresponding theorems.

Definition 4.1. *We shall say that the function $L(v)$ has the doubling property if there is a $C_L > 1$ such that*

$$L(2v) \le C_L L(v) \qquad (4.2)$$

for all $v > 0$. Also, we will say that L has the anti-doubling property if there are $A > 1$, $\varepsilon > 0$ such that

$$L(Av) \ge (1 + \varepsilon) L(v) \qquad (4.3)$$

for all $v > 0$.

Proposition 4.1. *If a graph satisfies (p_0) and there is a real function L with properties (4.2),(4.3) an in addition, the inequality*

$$\rho(B, A^c) \le L\left(\mu(A)\right) \mu(B)^{-1} \tag{4.4}$$

holds for all finite sets $\emptyset \ne B \subset A \subset \Gamma$, then

$$\overline{E}(A) \le CL\left(\mu(A)\right). \tag{4.5}$$

Proof   Let us fix an arbitrary $m > 1$ constant. Let $k = \min(j : m^{j-1}\mu(x) < \mu(A) \le m^j \mu(x))$ and let A_j denote the super-level set of $G^A(x,.)$ of volume of $\mu(A_j)$:

$$m^{j-1}\mu(x) < \mu(A_j) \le m^j \mu(x)$$

for $1 < j < k$, in particular let $A_0 = \{x\}$.

$$E_x(A) = \sum_{y \in A} G^A(x, y) = \sum_{y \in A} G^A(y, x) \frac{\mu(y)}{\mu(x)} \tag{4.6}$$

$$= \sum_{i=1}^{k} \sum_{y \in A_i \setminus A_{i-1}} G^A(y, x) \frac{\mu(y)}{\mu(x)} + G^A(x, x).$$

Now we recall that

$$G^A(y, x) \frac{1}{\mu(x)} = \rho(\Upsilon_y^r, A^c),$$

where Υ_y^r is the equipotential surface of y (cf. the refined model in Definition 2.30). Since A_{i-1} is a super-level set, Υ_y^r runs outside of A_{i-1} for $y \in A_i \setminus A_{i-1}$ and we get

$$G^A(y, x) \frac{1}{\mu(x)} = \rho(\Upsilon_y^r, A^c) \le \rho(A_{i-1}, A^c). \tag{4.7}$$

Our estimates can be continued as follows:

$$E_x(A) \le \sum_{i=1}^{k} \rho(A_{i-1}, A^c)\left(\mu(A_i) - \mu(A_{i-1})\right) + \mu(x)\rho(\{x\}, A^c) \tag{4.8}$$

and "integration by part" yields

$$E_x(A) \le \sum_{i=1}^{k} \rho(A_{i-1}, A^c)\left(\mu(A_i) - \mu(A_{i-1})\right) + \mu(x)\rho(\{x\}, A^c)$$

$$\le \sum_{i=1}^{k} \left(\rho(A_{i-1}, A^c) - \rho(A_i, A^c)\right)\mu(A_i) + \mu(x)\rho(\{x\}, A^c)$$

$$\le \sum_{i=1}^{k} \rho(A_{i-1}, A_i)\mu(A_i) + \mu(x)\rho(\{x\}, A^c).$$

Finally we can use condition (4.4) to receive

$$E_x(A) \le C \sum_{i=1}^{k} \frac{L(\mu(A_i))}{\mu(A_{i-1})} \mu(A_i) + \mu(x) \rho(\{x\}, A^c) \le C \sum_{i=0}^{k} L(\mu(A_i))$$

$$= C \sum_{i=0}^{k} L(m^i \mu(x)) \overset{(4.3)}{\le} CL(m^k \mu(x)) \overset{(4.2)}{\le} CL(\mu(A)),$$

which proves the statement. ∎

Theorem 4.2. *Assume that* (Γ, μ) *satisfies* (p_0) *and* L *has properties* (4.2) *and* (4.3), *then the following inequalities are equivalent*

$$\overline{E}(A) \le CL(\mu(A)) \tag{4.9}$$

for all finite $A \subset \Gamma$,

$$\lambda(A)^{-1} \le CL(\mu(A)) \tag{4.10}$$

for all finite $A \subset \Gamma$,

$$\|u\|_2 L \left(c' \frac{\|u\|_1}{\|u\|_2^2} \right) \le C \|\nabla u\|_2 \tag{4.11}$$

for all $u \in c_0(\Gamma)$,

$$\rho(B, A^c) \le CL(\mu(A)) \mu(B)^{-1} \tag{4.12}$$

for all finite $\emptyset \ne B \subset A \subset \Gamma$,

$$p_n(x, x) \le C \frac{1}{f(n)}, \tag{4.13}$$

where $f(t)$ *is the solution of the equation*

$$k = \int_0^{f(k)} \frac{L(S)}{s} ds.$$

Remark 4.1. This result is a slight generalization of Mathieu's work [75], developed for continuous time and for polynomial function $L(v) = v^\delta, \delta > 1$.

Proof   The implications can be seen as indicated below.

$$(4.9) \overset{(2.20)}{\Longrightarrow} (4.10) \overset{(3.8)}{\Longrightarrow} (4.12) \overset{\text{Proposition 4.1}}{\Longrightarrow} (4.9).$$

The other loop containing (4.10), (4.13), (4.11) is well known. (cf. [27]). The proof for the case $f(n) = n^{p/2}$ will be given in the Chapter 5 for demonstration purposes. ∎

4.2 Transient graphs

In this section we deal with transient random walks. With the aid of the Green's function we can formulate some results in the spirit of Carron ([19]).

Let us introduce some notations. The sequence of super-level sets is defined for an arbitrary but fixed $q \in (0, p_0^2)$ as

$$H_{x,i} = H_{x,q^i} = \{ y \in \Gamma : G(x,y) \geq q^i G(x,x) \}$$

and $a = a(q, x, \mu(A))$ is chosen to satisfy $\mu(H_{x,q^{a-1}}) < \mu(A) \leq \mu(H_{x,q^a})$.

Definition 4.2. *The resistance between a finite set A and "infinity" is defined for any reference point $x \in A$ as*

$$\rho_x(A) = \lim_{t \to 0} \rho(A, \Gamma \backslash H_{x,t}^r), \tag{4.14}$$

where

$$H_{x,t}^r = \{ w \in W : w = \overrightarrow{(y,z)} \times \alpha : G(y,x)(1-\alpha) + G(z,x)\alpha \geq tG(x,x) \}.$$

The main result of this chapter is the following general isoperimetric inequality.

Theorem 4.3. *For all transient weighted graphs satisfying (p_0),*

$$E_x(A) \leq C E_x(H_{a+1}) \tag{4.15}$$

where $C \geq 1$. independent of A and x.

This result can be interpreted as a weak isoperimetric inequality, showing the extremal property of super-level sets of the Green function. On this basis we shall see that it is enough to verify one of the equivalent statements in Theorem 4.2 for super-level sets of the Green function.

Lemma 4.1. *For all random walks and for sets $H \subset A \subset \Gamma$, where H is a super-level set of $G^A(x,y)$*

$$E_x(A) \geq \rho(H, A^c)\mu(H).$$

Proof   The proof is straightforward from the definition of super-level sets and from the observation that their boundaries are equipotential surfaces in the refined model. Let us consider

$$\min_{y \in H} \frac{1}{\mu(x)} G^A(y,x) =: \frac{1}{\mu(x)} G^A(\underline{y}, x) = \rho(\Upsilon_{\underline{y}}^r, A^c) \geq \rho(H, A^c)$$

$$E_x(A) = \sum_{y \in A} G^A(x,y) \geq \sum_{y \in H} \frac{\mu(y)}{\mu(x)} G^A(y,x)$$

$$\geq \sum_{y \in H} \mu(y)\rho(H, A^c).$$

∎

As a particular consequence of Lemma 4.1, we obtain the following.

Lemma 4.2. *For transient weighted graphs satisfying* (p_0),

$$E_x(H_{x,a+1}) \geq cq^{a+1}g(x,x)\mu(H_a), \tag{4.16}$$

where $q \in (0, p_0^2)$ *and* $c = \left(\frac{p_0}{q} - \frac{1}{p_0} \right)$.

Proof   For sake of simplicity, we suppress x in the subscript of $H_i = H_{x,i}$. From Lemma 4.1 we have

$$E_x(H_{a+1}) \geq \rho(H_a, H_{a+1}^c)\mu(H_a),$$

hence we need an estimate of $\rho(H_a, H_{a+1}^c)$. The interpolation of the refined network implies that

$$\rho(H_a, H_{a+1}^c) \geq \rho(H_a^r, \Gamma \backslash H_{a+1}^r),$$

and by shorting the equipotential surface of H_{a+1}^r the resistances are unchanged, hence we can calculate the resistance by using

$$\rho(H_a^r, \Gamma \backslash H_{a+1}^r) = \rho(H_a^r, \Gamma \backslash H_N^r) - \rho(H_{a+1}^r, \Gamma \backslash H_N^r).$$

If $N \to \infty$, thanks to monotonicity we have convergence and we obtain

$$\rho(H_a^r, \Gamma \backslash H_{a+1}^r) = \rho_x(H_a^r) - \rho_x(H_{a+1}^r). \tag{4.17}$$

We treat the new terms separately. For all $y \in H_a, s \sim y, s \notin H_a$

$$\mu(x)\rho_x(H_a^r) \geq \mu(x)\rho_x(\Upsilon_s^r) = G(s,x) \overset{(p_0)}{\geq} p_0 G(y,x) \geq p_0 q^a G(x,x), \tag{4.18}$$

on the other hand for $z \in H_{a+1}, v \notin H_{a+1}, z \sim v$

$$\mu(x)\rho_x(H_{a+1}^r) \leq \mu(x)\rho_x(\Upsilon_z^r) = G(z,x) \overset{(p_0)}{\leq} \frac{1}{p_0}G(v,x) < \frac{q^{a+1}}{p_0}G(x,x). \tag{4.19}$$

From estimates (4.18) and (4.19) we can receive the lower estimate

$$\rho(H_a^r, \Gamma \backslash H_{a+1}^r) \geq p_0 q^a g(x,x) - \frac{q^{a+1}}{p_0}g(x,x) \tag{4.20}$$

$$\geq \left(\frac{p_0}{q} - \frac{1}{p_0} \right) q^{a+1}G(x,x),$$

which results in the statement if $q < p_0^2$.   ∎

Proof [of Theorem 4.3] Let $a_i = \mu\left(A \cap (H_{x,i}\backslash H_{x,i-1})\right)$ and $h_i = \mu(H_{x,i}), i \geq 1$, $a_0 = \mu(x)$. The exit time from A can be decomposed by using the sets $A_i = A \cap H_{x,i}\backslash H_{x,i-1}$ as in (4.6) to receive

$$E_x(A) \leq G(x,x) \sum_{i=0}^{\infty} q^{i-1} a_i. \qquad (4.21)$$

Let us define the integer a for which $\mu(H_{a-1}) \leq \mu(A) < \mu(H_a)$ holds, furthermore let

$$l_i = h_i - h_{i-1} \text{ if } 1 \leq i \leq a$$
$$l_0 = h_0 \text{ and}$$
$$l_i = 0 \text{ if } i \geq a+1.$$

We can see that $l_i \geq a_i$ if $0 \leq i \leq a$ and consequently

$$\sum_{i=0}^{\infty} a_i \leq \mu(H_a) = h_a = \sum_{i=0}^{a} l_i.$$

As in the proof of Lemma 4.2 by using the potential surface of $y \in H_i\backslash H_{i-1}$ we have

$$\frac{1}{\mu(x)} G^{H_{a+1}}(y,x) = \rho(\Upsilon_y^r, H_{a+1}^c) \geq \rho(\Upsilon_y^r, \Gamma\backslash H_{a+1}^r) \qquad (4.22)$$
$$= \rho(\Upsilon_y^r, \Gamma\backslash H_N^r) - \rho(H_{a+1}^r, \Gamma\backslash H_N^r),$$

where $N > a+1$, and if $N \to \infty$, we obtain

$$\frac{1}{\mu(x)} G^{H_{a+1}}(y,x) \geq \rho(\Upsilon_y^r) - \rho\left(H_{a+1}^r\right).$$

Exactly the same arguments which have led to (4.20) can be used to derive that

$$\frac{1}{\mu(x)} G^{H_{a+1}}(y,x) \geq \rho_x(\Upsilon_y^r) - \rho_x\left(H_{a+1}^r\right)$$
$$\geq \frac{1}{\mu(x)}\left(q^i G(x,x) - \frac{1}{p_0} q^{a+1} G(x,x)\right).$$

The mean exit time can be decomposed as follows.

$$\frac{1}{G(x,x)} E_x(H_{a+1}) \geq \sum_{i=1}^{a} \min_{y \in H_i\backslash H_{i-1}} \frac{G^{H_{a+1}}(y,x)}{G(x,x)} \left(\mu(H_i) - \mu(H_{i-1})\right)$$
$$\geq \sum_{i=1}^{a}\left(q^i - \frac{1}{p_0}q^{a+1}\right) l_i = q\sum_{i=1}^{a} l_i q^{i-1} - \frac{1}{p_0} q^{a+1}\mu(H_a),$$

which, by Lemma 4.2, means that

$$\frac{C}{G(x,x)} E_x(H_{a+1}) \geq \sum_{i=0}^{a} l_i q^{i-1}. \tag{4.23}$$

The next step is to combine (4.23) and (4.21) to receive

$$\frac{1}{G(x,x)} \left(C E_x(H_{a+1}) - E_x(A) \right) \geq \sum_{i=0}^{\infty} \left(C l_i - a_i \right) q^{i-1},$$

which can be estimated as follows.

$$\sum_{i=0}^{\infty} \left(C l_i - a_i \right) q^{i-1} = \sum_{i=0}^{\infty} \left((C-1) l_i + l_i - a_i \right) q^{i-1} =$$

$$= \sum_{i=0}^{\infty} (C-1) l_i q^{i-1} + \sum_{i=0}^{a} \left(l_i - a_i \right) q^{i-1} + \sum_{i=a+1}^{\infty} \left(l_i - a_i \right) q^{i-1}$$

$$\geq \sum_{i=0}^{a} (C-1) l_i q^{i-1} - \sum_{i=a+1}^{\infty} a_i q^{i-1}$$

$$\geq (C-1) q^{a-1} \mu(H_a) - q^a \mu(A)$$

$$\geq q^a \mu(A) \left[q^{-1} (C-1) - 1 \right] \geq 0.$$

If we choose $C \geq \frac{q+1}{q} + \varepsilon, \varepsilon > 0$, then

$$C' E_x(H_{a+1}) \geq E_x(A).$$

■

The same arguments lead to the following less concise but useful conclusion.

Proposition 4.2. *For all transient weighted graphs with property* (p_0),

$$E_x(A) \leq C E_x(M_{k+1}) + C' \rho(M_{k+1}) \mu(M_k), \tag{4.24}$$

where $C \geq 1$. *is independent of A, and k is chosen to satisfy* $\mu(M_{x,k-1}) < \mu(A) \leq \mu(M_{x,k})$, *and* $M_{x,i}$s *are the largest possible super-level sets of* $G(x,y)$ *of volume* $\mu(H_{x,i}) \leq \mu(x) m^i$.

The following theorem is a slight generalization of Theorem 4.2. It is based on assumptions only on super-level sets of $G(x,y)$. Let $\mathcal{H}$ denote the super-level sets of $G(x,y)$.

Theorem 4.4. *Assume that* (Γ, μ) *is transient, satisfies* (p_0) *and L has property* (4.2) *with* $C_L < 2$ *and* (4.3), *then the following inequalities are equivalent:*

$$\overline{E}(H) \leq CL\left(\mu(H)\right) \text{ for all } H \in \mathcal{H}, \tag{4.25}$$

$$\lambda(H)^{-1} \le CL\left(\mu(H)\right) \ \textit{for all } H \in \mathcal{H}, \tag{4.26}$$

$$\rho(H) \le CL\left(\mu(H)\right)\mu(H)^{-1} \ \textit{for all } H \in \mathcal{H}, \tag{4.27}$$

$$p_n(x,x) \le C\frac{1}{f(n)}, \tag{4.28}$$

where $f(k)$ is the solution of

$$k = \int_0^{f(k)} \frac{L(S)}{s}\,ds.$$

Remark 4.2. Let us remark here that the inequality (4.27) is a generalization of Carron's [19]:

$$\mu(H) \le \gamma^{-p}, \tag{4.29}$$

where $\gamma = \rho(H)$. If we substitute $L(v) = v^{1-\frac{1}{p}}$ in (4.27), the inequality (4.29) follows.

Remark 4.3. The role of C_L needs some clarification. We can easily see from (4.28) that for $f(n) \simeq n^\gamma$

$$\sum_{k=1}^{k=n} p_k(x,x) \le C \sum_{k=1}^{k=n} k^{-\gamma}$$

if $\gamma = (\log_2 C_L)^{-1}$ and consequently the random walk is transient if $C_L < 2$. The reverse implication cannot be derived from (4.28) or from the equivalent conditions. In addition the first three conditions are based on the existence of super-level sets, hence the prior assumption of transience cannot be dropped.

Proof   [of Theorem 4.4] This proof is very similar to the previous ones. The inequality (4.26) follows from (4.25) by (3.8), and (4.25) follows from (4.27) just as in Proposition 4.1. To close the circle we have to show (4.26) $\Longrightarrow$ (4.27). For any $x \in \Gamma$ consider $G(x,y)$ and the super-level sets, in particular, let $M_i = M_{x,i}$ denote the set for which $2^{i-1}\mu(H) < \mu(M_i) \le 2^i\mu(H)$. Here we can refer again to the refined wire model and to the property of potential level-sets. If we apply this property to the surface of $M_i^r s$ it follows that

$$\rho(H) = \rho(H, \Gamma\backslash M_1^r) + \sum_{i=1}^{\infty} \rho(M_i^r, \Gamma\backslash M_{i+1}^r).$$

In order to return to non-refined subsets, we use comparison of the values of Green functions of neighboring vertices at the inner and outern boundary. Consider (4.18) and (4.19) to obtain

$$\rho(M_i^r, \Gamma\backslash M_{i+1}^r) \le C\rho(M_i, M_{i+1}^c).$$

From (3.8), it follows that

$$\rho(M_i, M_{i+1}^c) \leq \frac{1}{\lambda\,(M_{i+1})\,\mu\,(M_i)}$$

$$\overset{(4.26)}{\leq} C\frac{L(\mu\,(H_{i+1}))}{2^i\mu(H)} \leq C\frac{C_L^{i+1}L(\mu(H))}{2^i\mu(H)}.$$

Finally, by using the condition on C_L, we receive

$$\rho(H) \leq C\sum_{i=0}^{\infty}\frac{C_L^{i+1}L(\mu(H))}{2^i\mu(H)} \leq C\frac{L\,(\mu\,(H))}{\mu\,(H)}. \tag{4.30}$$

The link to the diagonal upper estimate is provided by Proposition 4.2 and we apply (4.30) to M_k.

$$\begin{aligned}
E_x(A) &\leq CE_x(M_{k+1}) + C'\rho(M_{k+1})\mu(M_k) \\
&\leq CL(\mu(M_{k+1})) \leq CL(\mu(M_k)) \\
&\leq CL(\mu(A))
\end{aligned}$$

Now we can apply Theorem 4.2 to get (4.28) and all the equivalent statements there as well. Of course, the return route from (4.28) via (4.9) to (4.25) is trivial, hence the proof is complete. ∎

Let us observe a weakness of Theorem 4.3. The volume growth from H_a to H_{a+1} is not controlled. If some kind of volume doubling condition holds for super-level sets, we can obtain a stronger estimate and all the usual isoperimetric optimum follows.

Definition 4.3. *The volume regularity property holds for super-level sets if there is a $C_H > 1$ and $q < q_0 < 1$ such that for all $x \in \Gamma$ and for all potential value γ*

$$\mu\,(H_{x,q\gamma}) \leq C_H\mu\,(H_{x,\gamma}), \tag{4.31}$$

and if there is a $c < q$ such that

$$\mu\,(H_{x,\gamma}) \leq c\mu\,(H_{x,q\gamma}). \tag{4.32}$$

Corollary 4.1. *For all transient weighted graphs satisfying (p_0) and* (4.31), (4.32)

$$E_x(A) \leq CE_x(H_a), \tag{4.33}$$

where $C \geq 1$. independent of $A \subset \Gamma$,

$$\rho\,(A) \leq C\rho\,(H_a) \tag{4.34}$$

and

$$\lambda^{-1}\,(A) \leq C\lambda^{-1}\,(H_a). \tag{4.35}$$

Proof   It is clear that the proof of Theorem 4.3 can be rephrased to get (4.33). First Lemma 4.2 can be reformulated to get for all $a \geq 0$

$$E_x \left(H_{x,a} \right) \geq cq^a G \left(x, x \right) \mu \left(H_{x,a} \right)$$

by choosing $q_0 = p_0^2$. In the main part of the proof H_{a+1} can be replaced by H_a and the result follows. The third inequality (4.35) follows from (4.33) by using the second volume regularity condition (4.32). Let μ_i denote $\mu \left(H_i \right)$ and observe that

$$\lambda^{-1} \left(A \right) \leq \overline{E} \left(A \right) \leq C E_{\overline{y}} \left(H_{\overline{y},a} \right) \leq C \sum_{i=1}^{a} \rho \left(H_{i-1}, H_i \right) \left[\mu_i - \mu_{i-1} \right],$$

where for shorter notation we suppressed $\overline{y}$. The next step is to use (4.32) :

$$C \sum_{i=1}^{a} \rho \left(H_{i-1}, H_a^c \right) \left[\mu_i - \mu_{i-1} \right]$$

$$\leq C \sum_{i=1}^{a} \rho \left(H_{i-1}, H_i^c \right) \mu_i$$

$$\leq C \sum_{i=1}^{a} q^{i-a} q^a c^{a-i} \mu_{a-1} \leq C q^a \mu_{a-1} \sum_{i=0}^{a} \left(\frac{c}{q} \right)^j$$

$$\leq C \rho \left(H_{a-1}, H_a^c \right) \mu \left(H_{a-1} \right) \leq C \lambda^{-1} \left(H_a \right).$$

Let us observe that we have also deduced that

$$E \left(H_a \right) \leq C \rho \left(H_{a-1}, H_a^c \right) \mu \left(H_a \right). \tag{4.36}$$

This can be used to obtain (4.34). We consider the level sets of the harmonic function h, set $h = 1$ on A and $h = 0$ on A^c and let A_i the level set of volume $Q^i \mu \left(A \right)$, where Q is chosen to satisfy $Q > C_H^2$. We can see that

$$\rho \left(A \right) = \sum_{i=0}^{\infty} \rho \left(A_i, A_{i+1}^c \right) \leq C \sum_{i=0}^{\infty} \frac{E \left(A_{i+1} \right)}{\mu \left(A_i \right)} \leq C \sum_{i=0}^{\infty} \frac{E \left(H_{A_{i+1}} \right)}{\mu \left(H_{A_{i+1}} \right)}.$$

From the observation (4.36) and by using twice (4.31) it follows that

$$\rho \left(A \right) \leq C \sum_{i=0}^{\infty} \frac{E \left(H_{A_{i+1}} \right)}{\mu \left(H_{A_{i+1}} \right)} \leq C \sum_{i=0}^{\infty} \rho \left(H_{a(A_{i+1}-1)}, H_{a(A_{i+1})}^c \right)$$

$$\leq C \sum_{i=0}^{\infty} \rho \left(H_{a(A_i)}, H_{a(A_{i+1})}^c \right) = C \rho \left(H_a \right)$$

■

In Section 8.9 we shall see that isoperimetric inequalities, similar to those studied in this section, can be obtained for recurrent graphs as well. For that we have to develop local or relative versions of inequalities. Such a generalization with respect to the volume is done by Coulhon and Grigor'yan [27].

4.3 Open problems

1. In Theorem 4.3 the super-level set H_k for $k = a + 1$ is used. The extra $+1$ is eliminated by using the volume regularity conditions (4.2, 4.3). The ideal result would be
$$E_x(A) \leq E_x\left(H^r_{x,a}\right),$$
where $\mu\left(H^r_{x,a}\right) = \mu(A)$. Is it possible to obtain such a result?

Let us put the problem in a more appealing form – the problem of "The Brownian sheep". We enclose a one acre part of a pasture with electric fence. Our sheep does Brownian motion in the inhomogeneous field (or does random walk on a weighted graph). We use an electric fence and want to maximize the expected time the electricity hits the sheep for the first time.

5

Polynomial volume growth

In this chapter we present a result for transient graphs with polynomial volume growth as an introduction. Its most interesting part is the proof that the Faber-Krahn inequality implies $(DUE_{\alpha,\beta})$, the diagonal upper estimate:

$$p_n(x,x) \leq n^{-\alpha/\beta}. \tag{5.1}$$

This is one of the key elements of the proof of two-sided sub-Gaussian estimates $(GE_{\alpha,\beta})$

$$Cn^{-\frac{\alpha}{\beta}} \exp\left[-\left(\frac{d^\beta(x,y)}{Cn}\right)^{\frac{1}{\beta-1}}\right] \leq \widetilde{p}_n(x,y) \leq Cn^{-\frac{\alpha}{\beta}} \exp\left[-\left(\frac{d^\beta(x,y)}{Cn}\right)^{\frac{1}{\beta-1}}\right]. \tag{5.2}$$

Here the modified heat kernel $\widetilde{p}_n = p_n + p_{n+1}$ is introduced to avoid parity problems. The proof is interesting on its own, since it goes through the Nash inequality which is equivalent to diagonal upper estimates for recurrent and transient graphs as well.

Theorem 5.1. ([48]) Let $\alpha > \beta > 1$. For any infinite connected weighted graph (Γ, μ) satisfying (p_0), the following equivalence holds:

$$(V_\alpha) + (G_{\alpha-\beta}) \iff (GE_{\alpha,\beta}),$$

where (V_α) stands for

$$V(x,R) \simeq R^\alpha, \tag{5.3}$$

and $(G_{\alpha-\beta})$ for

$$g(x,y) \simeq d(x,y)^{\beta-\alpha}. \tag{5.4}$$

Let us mention that this was the first characterization of weighted graphs which have two-sided sub-Gaussian estimates. Several authors obtained similar results on particular fractals. Here we do not present the whole proof. The steps not given here will be presented in a more general setup later on.

Diagonal upper estimates have been investigated by a number of papers over the past two or three decades. Let us mention here only a recent one dealing with the case of polynomial volume growth.

Theorem 5.2. *(Barlow, Coulhon, Grigor'yan [10])For a graph (Γ, μ) the following implications hold.*

$$V(x, R) \leq CR^\alpha \implies p_{2n}(x, x) \geq \frac{c}{V(x, \sqrt{Cn \log n})},$$

$$V(x, R) \geq cR^\alpha \implies p_{2n}(x, x) \leq Cn^{-\frac{\alpha}{\alpha+1}}$$

and there are examples of graphs (for all α) with polynomial growth $V(x, R) \simeq R^\alpha$, for which

$$p_{2n}(x, x) \simeq Cn^{-\frac{\alpha}{\alpha+1}}.$$

5.1 Faber-Krahn inequality and on-diagonal upper bounds

Let us recall that *a Faber-Krahn inequality (FK_ν) holds on (Γ, μ)* if there are constants $c > 0$ and $\nu > 0$ such that for all non-empty finite sets $A \subset \Gamma$,

$$\lambda_1(A) \geq c\mu(A)^{-1/\nu} \tag{5.5}$$

For example, (FK_ν) holds in $\mathbb{Z}^d$ with $\nu = d/2$.

Note that the parameter ν of the Faber-Krahn inequality is related to parameters α, β in the case of polynomial growth:

$$\nu = \frac{\alpha}{\beta}.$$

Theorem 5.3. *Let (Γ, μ) be a transient graph which satisfies (p_0), and let ν be a positive number. Then the following conditions are equivalent:*

(a) The Faber-Krahn inequality (FK_ν) holds.
(b) The on-diagonal heat kernel upper bound (DUE_ν), for all $x \in \Gamma$ and $n > 0$,

$$p_n(x, x) \leq Cn^{-\nu} \tag{5.6}$$

holds.
(3) An estimate of the volume of a level set of the Green kernel, for all $x \in \Gamma$ and $t > 0$,

$$\mu\{y : g(x, y) > t\} \leq Ct^{-\frac{\nu}{\nu-1}} \tag{5.7}$$

holds, provided $\nu > 1$.

The analogue of Theorem 5.3 for manifolds was proved by Carron [19]. The equivalence $(a) \Longleftrightarrow (b)$ was also proved in [43] for heat kernels on manifolds, and in [26, Proposition V.1] for random walks satisfying, also the condition $\inf_x P(x, x) > 0$.

We will provide detailed proof only for the implications $(a) \Longrightarrow (b)$ and $(c) \Longrightarrow (a)$. The implication $(b) \Longrightarrow (c)$ can be proved in the following way. By the theorem of Varopoulos [101], (DUE_ν) implies (4.1), the Sobolev inequality. Then we apply the argument of [19, Proposition 1.14] (adapted to the discrete setting) to show that (5.7) follows from the Sobolev inequality (4.1).

Let us note that our proof of $(a) \Longrightarrow (b)$ is valid for any $\nu > 0$. If $\nu > 1$, then we could apply the approach of [19] using the Sobolev inequality (4.1) as an intermediate step between (a) and (b). Instead, we use a Nash type inequality (cf. (4.11) in Chapter 4) which will be obtained in the following lemma.

Lemma 5.1. *Let (Γ, μ) be a weighted graph (which is not necessarily connected). Assume that for any non-empty finite set $A \subset \Gamma$,*

$$\lambda_1(A) \geq \Lambda(\mu(A)), \tag{5.8}$$

where $\Lambda(\cdot)$ is a non-negative non-increasing function on $(0, \infty)$. Let $f(x)$ be a non-negative function on Γ with finite support. Write

$$\sum_{x \in \Gamma} f(x)\mu(x) = a \quad \text{and} \quad \sum_{x \in \Gamma} f^2(x)\mu(x) = b.$$

Then, for any $s > 0$,

$$\frac{1}{2} \sum_{x \sim y} (\nabla_{xy} f)^2 \mu_{xy} \geq (b - 2sa) \Lambda(s^{-1}a). \tag{5.9}$$

Proof   If $b - 2sa < 0$, then (5.9) trivially holds. So, in the sequel we can assume that

$$s \leq \frac{b}{2a}. \tag{5.10}$$

Since $b \leq a \max f$, (5.10) implies $s < \max f$ and therefore, the following set

$$A_s = \{x \in \Gamma : f(x) > s\}$$

is non-empty. Consider the function $g = (f - s)_+$. This function belongs to $c_0(A_s)$, whence by the variational definition (2.18) of the eigenvalue we obtain

$$\frac{1}{2} \sum_{x \sim y} (\nabla_{xy} g)^2 \mu_{xy} \geq \lambda_1(A_s) \sum_{x \in \Gamma} g^2(x)\mu(x). \tag{5.11}$$

Let us estimate all terms in (5.11) via f. We start with the obvious inequality

$$f^2 \leq (f - s)_+^2 + 2sf = g^2 + 2sf, \tag{5.12}$$

which holds for any $s \geq 0$. It implies

$$g^2 \geq f^2 - 2sf,$$

whence

$$\sum_{x \in \Gamma} g^2(x)\mu(x) \geq b - 2s\,a. \tag{5.13}$$

As it follows from definition of A_s,

$$\mu(A_s) \leq s^{-1}a$$

whence by (5.8),

$$\lambda_1(A_s) \geq \Lambda(\mu(A_s)) \geq \Lambda(s^{-1}a). \tag{5.14}$$

Clearly, we also have

$$\sum_{x \sim y} (\nabla_{xy}g)^2 \mu_{xy} \leq \sum_{x \sim y} (\nabla_{xy}f)^2 \mu_{xy}.$$

Combining this with (5.14), (5.13) and (5.11), we obtain (5.9).   ∎

We will apply Lemma 5.1 for the function $\Lambda(v) = cv^{-1/\nu}$. Choosing $s = \frac{b}{4a}$ in (5.9), we obtain

$$\frac{1}{2} \sum_{x \sim y} (\nabla_{xy}f)^2 \mu_{xy} \geq c\,a^{-2/\nu}b^{1+1/\nu}. \tag{5.15}$$

This is a discrete analogue of the well-known *Nash inequality*.

Proof   [of $(a) \Longrightarrow (b)$ in Theorem 5.3]

STEP 1. Let f be a non-negative function on Γ with finite support. Write

$$b = \sum_{x \in \Gamma} f^2(x)\mu(x) \quad \text{and} \quad b' = \sum_{x \in \Gamma} [Pf(x)]^2 \mu(x)$$

where P is the Markov kernel of (Γ, μ). We then have

$$b - b' = (f, f)_{L^2(\Gamma,\mu)} - (Pf, Pf)_{L^2(\Gamma,\mu)} = (f, (I - P_2)f)_{L^2(\Gamma,\mu)}.$$

Clearly $Q = P_2$ is also a Markov kernel on Γ, reversible with respect to μ, and it is associated with another structure of a weighted graph on the set Γ. (Γ^*, μ^*) denote this weighted graph. As a set, Γ^* coincides with Γ and the measures μ and μ^* on vertices are the same. On the other hand, points x, y are connected by an edge on Γ^* if there is a path of length 2 from x to y in Γ, and the weight μ_{xy}^* on edges of Γ^* is defined by

$$\mu_{xy}^* = Q(x, y)\mu(x).$$

Denote the Laplace operator of (Γ^*, μ^*) by Δ^*. Then $\Delta^* = P_2 - I$ and, by the Green formula (2.17),

$$b - b' = - \sum_{x \in \Gamma} f(x)\, \Delta^* f(x) \mu(x) = \frac{1}{2} \sum_{x,y \in \Gamma} (\nabla_{xy} f)^2 \mu_{xy}^*. \qquad (5.16)$$

STEP 2. If A is a non-empty finite subset of Γ, then [27, Lemma 4.3] says that

$$\lambda_1^*(A) \geq \lambda_1(\overline{A}) \qquad (5.17)$$

where λ_1^* refers to the first eigenvalue of $-\Delta_A^*$ (this is based on the variational property (2.18) and on the fact that all eigenvalues of $-\Delta_A$ are in the interval $[\lambda_1(A), 2 - \lambda_1(A)]$). Therefore, (FK_ν) for (Γ, μ) implies that

$$\lambda_1^*(A) \geq c\mu(\overline{A})^{-1/\nu}. \qquad (5.18)$$

By (p_0) and Proposition 2.1, we have

$$\mu(\overline{A}) \leq \sum_{x \in A} V(x, 2) \leq C \sum_{x \in \Gamma} \mu(x) = C\mu(A).$$

Therefore, (5.18) implies

$$\lambda_1^*(A) \geq c' \mu(A)^{-1/\nu}$$

so that the graph (Γ^*, μ^*) also satisfies (FK_ν). ∎

Remark 5.1. This is the only place where (p_0) is used to ensure that $\mu(\overline{A}) \leq C\mu(A)$. If this inequality holds for some other reason, then the rest of the proof goes in the same way.

STEP 3. Fix some $y \in \Gamma$ and write

$$f_n(x) = p_n(x, y)$$

and

$$b_n = \sum_{x \in \Gamma} f_n^2(x)\mu(x) = p_{2n}(y, y).$$

Then $f_{n+1} = Pf_n$ and by (5.16) we obtain

$$b_n - b_{n+1} = \frac{1}{2} \sum_{x,y \in \Gamma} (\nabla_{xy} f)^2 \mu_{xy}^*.$$

Since the graph (Γ^*, μ^*) satisfies (FK), Lemma 5.1 can be applied. Since

$$\sum_{x \in \Gamma} f_n(x)\mu(x) = \sum_{x \in \Gamma} P_n(x, y) = 1,$$

(5.15) yields

$$\frac{1}{2} \sum_{x,y \in \Gamma} (\nabla_{xy} f)^2 \mu_{xy}^* \geq c b_n^{1+1/\nu},$$

whence

$$b_n - b_{n+1} \geq cb_n^{1+1/\nu}. \tag{5.19}$$

In particular, we see that $b_n > b_{n+1}$.

Next we apply an elementary inequality

$$\nu(x - y) \geq \frac{x^\nu - y^\nu}{x^{\nu-1} + y^{\nu-1}}, \tag{5.20}$$

which is true for all $x > y > 0$ and $\nu > 0$. Taking $x = b_{n+1}^{-1/\nu}$ and $y = b_n^{-1/\nu}$, from (5.20) and (5.19) we deduce

$$\nu(b_{n+1}^{-1/\nu} - b_n^{-1/\nu}) \geq \frac{b_{n+1}^{-1} - b_n^{-1}}{b_{n+1}^{-(\nu-1)/\nu} + b_n^{-(\nu-1)/\nu}} = \frac{b_n - b_{n+1}}{b_{n+1}^{1/\nu} b_n + b_n^{1/\nu} b_{n+1}} \geq \frac{cb_n^{1+1/\nu}}{2b_n^{1+1/\nu}} = \frac{c}{2}.$$

whence

$$b_{n+1}^{-1/\nu} - b_n^{-1/\nu} \geq \frac{c}{2\nu} = \text{const.}$$

Summing up this inequality in n, we get $b_n^{-1/\nu} \geq cn$ and $b_n \leq Cn^{-\nu}$.

Since $b_n = p_{2n}(y, y)$, we have proved that for all $y \in \Gamma$ and $n \in \mathbb{N}$,

$$p_{2n}(y, y) \leq Cn^{-\nu}, \tag{5.21}$$

which is (5.6) for all *even* number of steps.

STEP 4. By the semigroup identity for any $0 < k < m$ we have,

$$p_m(x, y) = \sum_{z \in \Gamma} p_{m-k}(x, z) p_k(z, y) \mu(z). \tag{5.22}$$

In particular, if $m = 2n$, $k = n$ and $y = x$, then

$$p_{2n}(x, x) = \sum_{z \in \Gamma} p_n^2(x, z) \mu(z). \tag{5.23}$$

On the other hand, (5.22), the Cauchy–Schwarz inequality and (5.23) imply

$$p_{2n}(x, y) = \sum_{z \in \Gamma} p_n(x, z) p_n(z, y) \mu(z)$$

$$\leq \left[\sum_{z \in \Gamma} p_n^2(x, z) \mu(x) \right]^{\frac{1}{2}} \left[\sum_{z \in \Gamma} p_n^2(y, z) \mu(z) \right]^{\frac{1}{2}}$$

$$= p_{2n}(x, x)^{1/2} p_{2n}(y, y)^{1/2}. \tag{5.24}$$

Therefore, (5.21) yields, for all $x, y \in \Gamma$,

$$p_{2n}(x, y) \leq Cn^{-\nu}.$$

This implies (DUE_ν) also for *odd* times if we observe by (5.22) that

$$p_{2n+1}(x, y) = \sum_{z \in \Gamma} p_{2n}(x, z) p(z, y) \leq \max_{z \in \Gamma} p_{2n}(x, z). \tag{5.25}$$

Proof   [of $(c) \Rightarrow (a)$ in Theorem 5.3] From Lemma 2.3 we know that

$$\lambda_1(A) \geq \overline{E}(A)^{-1}. \tag{5.26}$$

On the other hand, for any $x \in A$,

$$E_x(A) = \sum_{y \in A} G^A(x, y) = \sum_{y \in A} g^A(x, y)\mu(y) = \int_0^\infty \mu\left\{g^A(x, \cdot) > t\right\} dt.$$

Let us fix some $t_0 > 0$ and estimate the integral above using (5.7), $g^A \leq g$ and the fact that

$$\mu\left\{g^A(x, \cdot) > t\right\} \leq \mu(A).$$

We obtain

$$\sum_{y \in A} G^A(x, y) \leq \int_0^{t_0} \mu(A)dt + \int_{t_0}^\infty Ct^{-\frac{\nu}{\nu-1}} dt = \mu(A)t_0 + Ct_0^{-\frac{1}{\nu-1}}.$$

Finally, we choose t_0 to equate the two terms on the right-hand side, that is,

$$t_0 \simeq \mu(A)^{-\frac{\nu-1}{\nu}},$$

whence

$$\sum_{y \in A} G^A(x, y) \leq C\mu(A)^{1/\nu}. \tag{5.27}$$

Finally, (5.27) and (5.26) imply (5.5).   ∎

Exercise 5.1. Develop the details of the proof of (5.17).

Part II

Local theory

6

Motivation of the local approach

In this chapter we provide some intuitive arguments to relax the conditions imposed on weighted graphs on which random walks take place. We develop a "local" framework, local in the sense that both the volume and the mean exit time of a ball depend on its center, so no spatial homogeneity is assumed. Let us summarize the conditions providing the framework.

We assume (p_0), that is, there is a $p_0 > 0$ such that for all edges $x \sim y$

$$P(x, y) \geq p_0.$$

We assume volume doubling: there is a C_V such that for all $x \in \Gamma, R > 0$

$$\frac{V(x, 2R)}{V(x, R)} \leq C_V,$$

and the time comparison principle: there is a $C > 0$ such that for all $x \in \Gamma, R > 0, y \in B(x, R)$

$$\frac{E(x, 2R)}{E(y, R)} < C.$$

First we review some fundamental properties of the exit time to indicate how the framework of the theory arises in a "natural " way, then we give an example.

6.1 Properties of the exit time

As it was indicated in the introduction, the mean exit time is a central subject in the study of random walks. The classical results were based on the same $E(x, R) \simeq R^2$ scaling and it became a general belief that in all properly behaving (non-hyperbolic) spaces (Ahlfors regular or amenable)

$$E(x, R) \geq cR^2$$

is true. The first lower estimates were given by Kesten and Kusuoka.

Theorem 6.1. *(See [61],[66]) If (Γ, μ) satisfies (V_α), then for $R \geq 1, x \in \Gamma$*

$$E(x, R) \geq \frac{cR^2}{(\log R)^{1/2}}.$$

Only in 1989 did Barlow and Perkins [13] give an example of a graph for which the lower bound is attained. Very recently Virág [104] proved that

$$E_x(\tau_y) \simeq \frac{d(x, y)^2}{\log(d(x, y))^{1/2}}$$

for graphs where $\mu(y) \leq \mu(x) d(x, y)^p$ holds for some $p > 0$. The author showed for a class of graphs in [94] that

$$2 \leq \beta \leq \alpha + 1, \tag{6.1}$$

which is revisited by Barlow in [3], also giving examples of graphs with all possible α and β satisfying the restriction (6.1). In this section we elaborate on the behavior of the exit time which helps to understand the basic behavior of random walks.

For the following result we define the local sub-Gaussian kernel function which replaces the Gaussian and sub-Gaussian exponents in (1.3) and (1.4) respectively.

Definition 6.1. *The local kernel function $R \wedge n \geq k = k(x, n, R) \geq 1$, with respect to a function $F(x, R)$, is defined as the maximal integer for which*

$$\frac{n}{k} \leq qF(x, \left\lfloor \frac{R}{k} \right\rfloor), \tag{6.2}$$

or $k = 1$ by definition if $n > qF(x, R)$ or there is no appropriate k. Here q is a small fixed constant which will be specified later.

Typically F will be the mean exit time, if not, it will be made clear.

Definition 6.2. *The local sub-Gaussian exponent $R \wedge n \geq k(n, R, A) \geq 1$ with respect to a function $F(x, R)$ for $A \subset \Gamma$, is the maximal integer k for which*

$$\frac{n}{k} \leq C \max_{z \in A} F(z, \frac{R}{k}). \tag{6.3}$$

Write $k_x(n, R) = k(n, R, B(x, R))$.

Definition 6.3. *Let $R \wedge n \geq l = l(x, n, R) \geq 1$ be the minimal integer, for which*

$$\frac{n}{l} \geq CF(x, \frac{R}{l}), \tag{6.4}$$

or $l = R$ by definition if there is no appropriate l. The constant C will be specified later.

Definition 6.4. *The local sub-Gaussian lower exponent* $l(n, R, A)$, *with respect to a function* $F(x, R)$ *for* $A \subset \Gamma$ *is the maximal integer* l *for which*

$$\frac{n}{l} \geq C \max_{z \in A} F(z, \frac{R}{l}). \tag{6.5}$$

We will choose $C > C_F \left(\frac{1}{\delta}\right)^\beta$ (see (7.20) and (10.23) below).

Definition 6.5. *The global sub-Gaussian exponent* $R \wedge n \geq m = m(n, R) \geq 1$ *is defined as the maximal integer for which*

$$\frac{n}{m} \leq q \inf_{y \in \Gamma} F(y, \frac{R}{m}), \tag{6.6}$$

or $m = 1$ *by definition if there is no appropriate* m.

Lemma 6.1. *Assume that* (TC) *holds and* $\beta > 1$. *Then for all* $x \in \Gamma, R, n > 0$

$$k_x(n, R) + 1 \geq c \left(\frac{E(x, R)}{n}\right)^{\frac{1}{\beta - 1}}, \tag{6.7}$$

and if $\beta' > 1$,

$$l(x, n, R) \leq C \left(\frac{E(x, R)}{n}\right)^{\frac{1}{\beta' - 1}} \tag{6.8}$$

Proof   The statement follows easily from (TC).   ∎

Theorem 6.2. *If* (p_0) *and* $\left(\overline{E}\right)$ *hold, then there is a* $c_0 > 0$ *such that*

$$P(T_{x,R} < \frac{1}{2}E(x, R)) < 1 - c_0 \tag{6.9}$$

for all $x \in \Gamma, R > 1$, *and* $DLE(E)$, *the local diagonal lower estimate*

$$p_{2n}(x, x) \geq \frac{c}{V(x, e(x, 2n))} \tag{6.10}$$

holds, where $e(x, .)$ *is the inverse of* $E(x, .)$ *in the second variable. Furthermore,*

$$\mathbb{P}(T_{x,R} < n) \leq C \exp\left[-ck_x(n, R)\right]. \tag{6.11}$$

This result immediately raises the question. How can we get a diagonal upper estimate $DUE(E)$

$$p_{2n}(x, x) \leq \frac{C}{V(x, e(x, 2n))} \tag{6.12}$$

for the heat kernel similar to (6.10)?

The answer to this question will be given within the framework of the local theory defined by the conditions $(p_0), (VD)$ and (TC). We prove Theorem 6.2 and then present some further results which are needed.

Definition 6.6. *The diagonal lower estimate (6.10) will be referred to as* $DLE\,(F)$ *for general functions* F.

The following two lemmas will help to prove Theorem 6.2. An alternative way to show (6.11) can be found in [48].

Corollary 6.1. *The inequality (6.9) implies that*

$$P(T_{x,R} > n) > c_0 \tag{6.13}$$

if $n \le \frac{1}{2}E\,(x,R)$.

Lemma 6.2. *Let* $T, \tau_1, \tau_2, ..., \tau_m$ *be non-negative random variables such that* $T \ge \sum_{k=1}^m \tau_i$. *Assume also that, for some* $a \in (0,1)$ *and* $b > 0$ *and all* $k = 1, 2, ..., m$ *and* $t \ge 0$,

$$\mathbb{P}\left(\tau_k \le t | \sigma(\tau_1, ... \tau_{k-1})\right) \le a + bt. \tag{6.14}$$

Then, for all $t \ge 0$,

$$\mathbb{P}(T \le t) \le \exp\left[2\left(\frac{btm}{a}\right)^{1/2} - m\log\frac{1}{a}\right]. \tag{6.15}$$

Proof   For a random variable η assume that $P(\eta < t) = (p + at) \wedge 1$, then

$$\mathbb{E}\left(e^{-\lambda \xi_i} | \sigma\left(\xi_1, ..\xi_{i-1}\right)\right) \le \mathbb{E}\left(e^{-\lambda \eta}\right)$$

$$= p + \int_0^{(1-p)a} e^{-\lambda t} a dt \le p + a\lambda^{-1}.$$

This implies that

$$\mathbb{P}\left(V \le t\right) = \mathbb{P}\left(e^{-\lambda V} \ge e^{-\lambda t}\right) \le e^{\lambda t} \mathbb{E}\left(e^{-\lambda V}\right)$$

$$e^{\lambda t} \mathbb{E}\left(\exp\left[\lambda \sum_{i=1}^n \xi_i\right]\right) \le e^{\lambda t}\left(p + a\lambda^{-1}\right)^n$$

$$\le p^n \exp\left(\lambda t + \frac{an}{\lambda p}\right).$$

The result is obtained by choosing $\lambda = (an/pt)^{1/2}$.   ∎

Lemma 6.3. *For all* $A \subset \Gamma$, $x \in A$ *and* $t \ge 0$ *we have that*

$$\mathbb{P}_x(T_A < t) \le 1 - \frac{E_x(A)}{2\overline{E}(A)} + \frac{t}{\overline{E}(A)}. \tag{6.16}$$

Proof   Write $n = \lfloor t \rfloor$ and observe that

$$T_A \le 2t + 1_{\{T_A > t\}} T_A \circ \theta_n,$$

where θ_n is the time shift operator. Since $\{T_A > t\} = \{T_A > n\}$, by the strong Markov property we obtain

$$\mathbb{E}_x(T_A) \le 2t + \mathbb{E}_x\left(1_{\{T_A > t\}} \mathbb{E}_{X_n}(T_A)\right) \le 2t + \mathbb{P}_x\left(T_A > t\right) 2\overline{E}(A).$$

By applying the definition $E_x(A) = \mathbb{E}_x(T_A)$, we obtain (6.16).   ∎

Proof [of Theorem 6.2] The first statement is immediate from Lemma 6.3 using $A = B(x, R), n = \frac{1}{2}E(x, R)$ and $(\overline{E})$.

$$\mathbb{P}_x(T_{x,R} < n) \le 1 - \frac{E(x, R)}{2\overline{E}(x, R)} + \frac{n}{\overline{E}(x, R)}$$

$$= 1 - \frac{E(x, RA)}{4CE(x, R)} \le 1 - c_0.$$

The *DLE* follows easily. Let us recall that

$$\mathbb{P}_x(T_{x,R} > n) \ge c_0 > 0$$

and proceed with

$$c_0^2 \le \mathbb{P}_w(T_{x,R} > n)^2 \le \left(e_w^* P_n^A \mathbf{1}\right)^2$$

$$\le \left(\sum_{y \in A} P_n^A(x, y) \sqrt{\frac{\mu(y)}{\mu(y)}}\right)^2 \le \left(\sum_{y \in A} \mu(y)\right) \left(\sum_{y \in A} \frac{P_n^A(x, y)^2}{\mu(y)}\right) \quad (6.17)$$

$$= \mu(A) \left(\sum_{y \in A} P_n^A(x, y) \frac{P_n^A(y, x)}{\mu(x)}\right) \le \frac{1}{\mu(x)} V(x, R) P_{2n}^A(x, x) \quad (6.18)$$

$$\le \frac{1}{\mu(x)} V(x, R) P_{2n}(x, x). \quad (6.19)$$

To get the statement let us finally recall that $R = e(x, 2n)$. To prove the next crucial statement (6.11), first we assume that $R \le n \le q \min_{y \in B(x,R)} E(y, R)$, otherwise the statement is trivial. Let us write $r = \lfloor \frac{R}{m} \rfloor$ and define a series of exit times recursively

$$\tau_1 = T_{x,r},$$
$$\tau_{k+1} = T_{y_k, r}, \text{ where } y_k = X_{\tau_k}, \text{ for } k = 1, ..., m.$$

Since $d(X_{\tau_{k+1}}, X_{\tau_k}) < r$ and $mr \le R$, we see that $d(X_n, x) < R$ for all $n \le \sum_{k=1}^m \tau_k$. Therefore, we obtain

$$T_{x,R} \ge \sum_{k=1}^m \tau_k. \quad (6.20)$$

Lemma 6.3 can be applied to $A = B(X_{\tau_k}, r)$, thus

$$\mathbb{P}(\tau_k \le t | \sigma(\tau_1, ...\tau_{k-1})) \le a + bn, \quad (6.21)$$

where

$$a = \sup_y \left(1 - \frac{E(y, r)}{2\overline{E}(y, r)}\right) \in [\frac{1}{2}, 1 - \varepsilon], \quad (6.22)$$

and

$$b = \frac{1}{2 \min\limits_{y \in B(x,R)} \overline{E}(y,r)} \leq \frac{1}{2 \min\limits_{y \in B(x,R)} E(y,r)}. \tag{6.23}$$

Having (6.20) and (6.21), we apply Lemma 6.2 which yields

$$\mathbb{P}_x(T_{x,R} \leq n) \leq \exp\left[2\left(\frac{bnm}{a}\right)^{1/2} - m \log \frac{1}{a}\right].$$

Rewriting the inequality (6.2) in the form

$$n \leq c_2 m \min_{y \in B(x,R)} E(y,r)$$

and combining it with (6.23), we obtain $bn \leq \frac{1}{2}c_2 m$. If $c_2 > 0$ is sufficiently small (say $c_2 = \frac{1}{16} \log^2 \frac{1}{1-\varepsilon}$), then together with $1/2 \leq a \leq 1 - \varepsilon$, it implies

$$2\left(\frac{bnm}{a}\right)^{1/2} - m \log \frac{1}{a} \leq -c_1 m,$$

(where $c_1 = \frac{1}{2} \log \frac{1}{1-\varepsilon}$) which completes the proof of (6.11).   ∎

Remark 6.1. Let us observe that we have proved a bit more than the $DLE\,(E)$ namely that $(\overline{E})$ implies

$$p_{2n}^{B(x,R)}(x,x) \geq \frac{c}{V(x, e(x, 2n))} \tag{6.24}$$

for $n < \varepsilon E$, where $R = e(x, 2n)$, or in another form using $E = E(x, R)$

$$p_E^{B(x,R)}(x,x) \geq \frac{c}{V(x, R)}. \tag{6.25}$$

6.2 Examples

In this section we give simple examples of a graphs satisfying (VD) and (TC). The first one is the weighted Vicsek tree introduced in Example 2.1. Her we discuss its properties in detail.

Example 6.1. It is clear that (VD) is satisfied on (Γ, μ). Due to the tree structure of G, it is easy to compute the Green kernel $g^{\Gamma_k}(x_k, \cdot)$. Let z_k be the only point in $\overline{G_k} \setminus G_k$ (see Figure 6.1).
    Let $g_k(y)$ be a function on Γ satisfying the following conditions:

- g_k vanishes outside Γ_k, in particular at z_k;

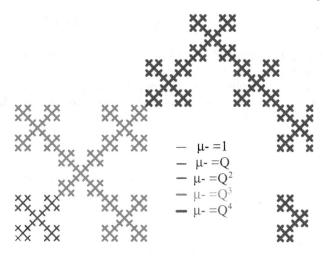

$$- \quad \mu\text{-}=1$$
$$- \quad \mu\text{-}=Q$$
$$- \quad \mu\text{-}=Q^2$$
$$- \quad \mu\text{-}=Q^3$$
$$- \quad \mu\text{-}=Q^4$$

Fig. 6.1. The weighted Vicsek tree

- g_k increases linearly along the path from z_k to x_k with a constant increment c_k at each step;
- g_k remains constant along all other paths in Γ_k.

These conditions uniquely determine g_k up to c_ks. Clearly, g_k is harmonic in $G_k \setminus x_k$. At point x_k, we have $\Delta g_k = -c_k/4$. Therefore, if we take $c_k = 4/\mu(x_k) = a^{-k}$, then for all $y \in \Gamma_k$, we obtain

$$\Delta g_k(y) = -\delta_{x_k}(y),$$

hence $g_k \equiv g^{\Gamma_k}(x_k, \cdot)$.

For any point y on the paths from z_k to x_k, we have $g_k(y) = c_k d(y, z_k)$. In particular, for any $y \in B(x_k, \frac{1}{3}R_k)$,

$$g_k(y) \simeq \left(\frac{3}{a}\right)^k \quad \text{and} \quad g_k(y)\mu(y) \simeq 3^k.$$

Therefore, by (7.2)

$$E(x_k, R_k) = \sum_{y \in \Gamma_k} g_k(y)\mu(y) \simeq 3^k |\Gamma_k| \simeq 15^k \simeq R_k^\beta,$$

where $\beta = \log_3 15$. It is easy to show that the same relation $E(x, R) \simeq R^\beta$ holds for all $x \in \Gamma$ and $R \geq 1$ which proves (E_β).

The Green kernel $g_k = g^{\Gamma_k}(x_k, \cdot)$ constructed above is nearly radial. A similar argument shows that the same is true for all balls in Γ which implies (H).

Let G_i be a the sub-graph of the Vicsek tree which contains the root z_0 and has diameter $D_i = 23^i$. Let z_i denote the vertices on the infinite path with $d(z_0, z_i) = D_i$. Write $G'_i = (G_i \backslash G_{i-1}) \cup \{z_{i-1}\}$ for $i > 0$, the annuli defined by Gs.

The new graph is defined by stretching the Vicsek tree as follows. Consider the sub-graphs G'_i and replace each of their edges by a path of length $i + 1$. Denote the new sub-graphs by A_i, the new blocks by $\Gamma_i = \cup_{j=0}^i A_i$ and the new graph is $\Gamma = \cup_{j=0}^\infty A_j$. Denote the cut-point between A_i and A_{i-1} again by z_i. For $x \neq y, x \sim y$, let $\mu_{x,y} = 1$ (see Figure 6.2).

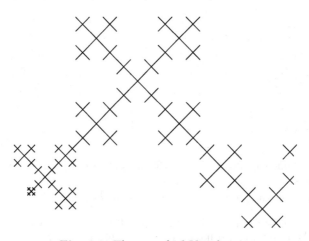

Fig. 6.2. The stretched Vicsek tree

We can see that neither the volume nor the mean exit time grows polynomially on Γ, on the other hand Γ is a tree and the resistance grows asymptotically linearly on it. We show that (VD) and (TC) hold on Γ.

Let us recognize some straightforward relations first:

$$d(z_0, z_n) = d(z_0, z_{n-1}) + 2n3^n < n3^{n+1} < (n+2)3^{n+1} = d_{n+1}, \qquad (6.26)$$

$$\mu(\Gamma_n) = C\left(4 + \sum_{i=1}^n 2(i+1)4^i\right) \simeq n4^n \simeq \mu(A_n),$$

$$\rho(\{x\}, B(x,R)^c) \simeq \rho(x, R, 2R) \simeq R,$$

$$E(x,R) \leq CRV(x,R).$$

Lemma 6.4. *The tree Γ satisfies (VD).*

Proof　Write $L_i = d(z_0, z_i), d_i = \frac{1}{2}(L_i - L_{i-1})$ and observe that $L_{n-1} \simeq L_n \simeq d_n$. Let us consider a ball $B(x, 2R)$ and an $N > 0$ so that $x \in A_N$ and a k such that

$$d_{k-1} \leq R < d_k.$$

First we assume that the ball is large relative to the position of the center. In this case the large scale properties of the graph are dominant.

Case 1. $k \geq N$.

For convenience we introduce a notation. Ω_n denote one of the blocks ($B_n...E_n$) of A_n of diameter d_n (see Figure 6.3). There is a block Ω_{k-2} which

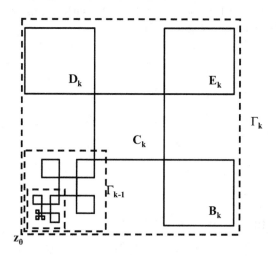

Fig. 6.3. The blocks of the Vicsek tree

contains x. It is clear that $\Omega_{k-2} \subset B(x, R)$ and

$$V(x, R) \geq \mu(\Omega_{k-2}) \simeq \mu(\Gamma_{k+1}).$$

On the other hand, $R < L_k$ which implies that $B(x, 2R) \subset \Gamma_{k+1}$ and (VD) follows from $\mu(\Gamma_{k+1}) \simeq \mu(\Omega_{k-2})$.

Case 1. $k < N$.

Now we have to separate sub-cases. Again let us fix that $x \in \Omega_N$. Write $d = d(x, z_{N-1})$. If x is not in the central block C_N of A_N, then $B(x, R) \subset A_N \cup A_{N+1}$ and since these parts of the graph contain only paths of a length of $N + 1$ or $N + 2$, volume doubling follows from the fact that it holds for the original Vicsek tree. The same applies if x is in the central block but $B(x, 2R) \subset A_N$. Finally, if $B(x, 2R) \cap \Gamma_{N-1} \neq \varnothing$, then $R \geq 2d_{N-1} > d_{N-1}$ which by the definition of k means that $k = N - 1, B(x, R) \supset \Omega_{N-1}$ and on the other hand, $B(x, 2R) \subset \Gamma_{N+1}$ which again gives (VD). ∎

The elliptic Harnack inequality follows from the fact that Green functions are nearly radial. Let us recall that

$$cp\left(x,R,2R\right)V\left(x,R\right) \le E\left(x,2R\right) \le Cp\left(\{x\},B\left(x,2R\right)^{c}\right)V\left(x,2R\right), \quad (6.27)$$

where the lower bound follows from the elliptic Harnack inequality, and the upper bound holds in general.

The following step is to show (TC). Observe that the graph has linear resistance growth which means that (6.27) and (VD) imply (TC). For balls centered in the root, $\beta > 2$ (see Figure 6.4) while for a vertex in the middle of a long path, far from the root $\beta = 2$ for balls with radius comparable with the length of the path (see Figure 6.5).

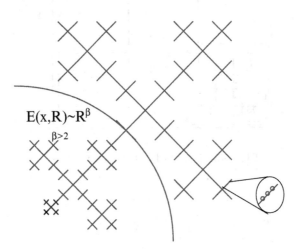

Fig. 6.4. Test ball around the root

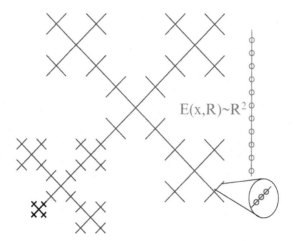

$E(x,R) \sim R^2$

Fig. 6.5. A test ball far from the root

7

Einstein relation

With this chapter we start a systematic foundation of the local theory. A key tool for that is the Einstein relation (**ER**) :

$$E\left(x, 2R\right) \simeq v\left(x, R, 2R\right) \rho\left(x, R, 2R\right). \tag{7.1}$$

In [94] it has been shown that

$$\beta = \alpha + \gamma$$

holds for a large class of graphs. See the history and comments below (1.5). A more detailed picture can be obtained by considering the resistance and volume growth properties. Recent studies (cf. [3], [5], [6], [48], [49], [98]) show that the relevant form of the Einstein relation is (7.1).

Our aim with this chapter is to give reasonable conditions for this relation and show some further properties of the mean exit time which are essential in the investigation of diffusion, in particular for obtaining heat kernel estimates.

Similar estimates for particular structures have been considered (cf. [3],[5], [54])

The Einstein relation provides a simple connection between the members of the triplet of mass, resistance and mean exit time. Examples show that without some natural restrictions any two of them are "independent" (cf. Lemma 5.1, 5.2 [3] and [31] and references there). From the point of view of physics, it seems natural to impose conditions on mass and resistance.

It is interesting to see that the conditions proved to be most natural for the Einstein relation are the ones that provide the basic framework for the study of heat kernel estimates.

The chapter is structured in the following way. In the consecutive sections we gradually change the set of conditions. In Section 7.1 we show the Einstein relation by using the time comparison principle. This is a condition on the mean exit time which might be challenged as an initial assumption in the study of diffusion. In order to eliminate this deficiency, in Section 7.2 we replace this condition by a pair of conditions which reflect resistance properties. The conditions will become increasingly restrictive to meet the requirements of heat kernel estimates. In particular, the strong assumption of the elliptic Harnack inequality will be used. Section 7.1,7.2 and 7.3 provide two more results on the Einstein relation under different conditions and discuss several further properties of the mean exit time.

7.1 Weakly homogeneous graphs

The mean exit time can be expressed by Green functions:

$$E(x, R) = \sum_{y \in B(x,R)} G^{B(x,R)}(x, y). \tag{7.2}$$

In the sequel this observation will be frequently used.

Theorem 7.1. *If* (p_0), (VD), (TC) *hold, then*

$$E(x, 2R) \simeq \rho(x, R, 2R)v(x, R, 2R) \simeq \lambda^{-1}(x, 2R). \tag{7.3}$$

Proof   We start with the general inequalities (3.9) and (2.20). We obtain

$$\rho(x, R, 2R)V(x, R) \tag{7.4}$$

$$\leq \lambda^{-1}(x, 2R)$$

$$\leq \overline{E}(x, 2R)$$

$$\leq CE(x, 2R),$$

where in the last step (TC) was used. For the upper estimate, let us use Lemma 3.3

$$\rho(x, R, 2R)V(x, 2R) \geq \min_{y \in \partial B(x, \frac{3}{2}R)} E(y, \frac{1}{2}R),$$

which by (TC) results in

$$\rho(x, R, 2R)v(x, R, 2R) \geq cE(x, 2R).$$

∎

The volume doubling property implies anti-doubling for volume, a similar implication holds for the mean exit time.

Definition 7.1. *We have the anti-doubling property* (aDT) *for* E *if there is an* $A_T \geq 2$ *such that*

$$E(x, A_T R) > 2E(x, R). \tag{7.5}$$

Proposition 7.1. *If* (wTC) *holds, then the anti-doubling property holds for* $E(x, R)$.

Proof   Consider $y \in S = \partial B(x, R)$ and observe that the iterated use of (wTC) implies

$$E(y, R) > cE(x, R).$$

It follows from the strong Markov property that

$$E(x, 2R) \geq E_x \left(T_{B(x,R)} + E_\xi (T_A) \right)$$
$$\geq E(x, R) + \min_{y \in S} E(y, R) \geq (1 + c) E(x, R),$$

where the short notations $\xi = X_{T_{B(x,R)}}$ and $A = B(\xi, R)$ were used for convenience. By iterating this procedure, we get the result for $A_E = 2^k$ with $k = \left\lceil \frac{\log 2}{\log(1+c)} \right\rceil$. ∎

Theorem 7.2. *If for a weighted graph (Γ, μ) the conditions $(p_0), (VD), (H)$ and (wTC) hold, then*

$$E(x, 2R) \simeq \rho(x, R, 2R) v(x, R, 2R).$$

Proposition 7.2. *If $(p_0), (wTC)$ hold for (Γ, μ), then there is a $C \geq 1$. such that for all $x \in \Gamma, R > 0$*

$$E(x, 2R) \leq C\rho(x, R, 5R) v(x, R, 5R).$$

Proof   Let us consider the annulus $D = B(x, 5R) \setminus B(x, R)$. We apply Corollary 3.6 to $A = B(x, R), B = B(x, 5R)$ to receive

$$\rho(x, R, 5R) v(x, R, 5R) \geq E_a(T_B).$$

Now we use the Markov property for the stopping time $T_{B(x,3R)}$. It is clear that a walk starting in $B(x, R)$ and leaving $B(x, 5R)$ should cross $\partial B(x, 3R)$. ξ denote the first hitting (random) point. It is again evident that the walk continuing from ξ should leave $B(\xi, 2R)$ before leaving $B(x, 5R)$. This means that

$$\rho(x, R, 5R) v(x, R, 5R)$$
$$\geq E_a(T_B) \geq \min_{y \in S(x,3R)} E(y, 2R)$$
$$\geq cE(x, 2R),$$

where the last inequality follows from the repeated use of (wTC). ∎

Proof   [of Theorem 7.2] The lower estimate is easy. Write $B = B(x, 2R)$. We know that (H) implies $wHG(2)$ and consequently

$$E(x, 2R) = \sum_{y \in B} G^B(x, y) \tag{7.6}$$

$$\geq \sum_{y \in B(x,R)} g^B(x, y) \mu(y)$$

$$\geq c\rho(x, R, 2R) V(x, R)$$

$$\geq c\rho(x, R, 2R) v(x, R, 2R).$$

The upper estimate uses the fact that the Harnack inequality implies the doubling property of resistance. From Proposition 7.2 we have that

$$E\left(x, 2R\right) \leq C\rho\left(x, R, 5R\right) v\left(x, R, 5R\right),$$

but (3.45) and volume doubling result in

$$E\left(x, 2R\right) \leq C\rho\left(x, R, 8R\right) v\left(x, R, 5R\right)$$
$$\leq C\rho\left(x, R, 2R\right) V\left(x, 5R\right)$$
$$\leq C\rho\left(x, R, 2R\right) v\left(x, R, 2R\right).$$

∎

The following lemma is weaker than the observation in Remark 3.11 but the proof is so easy that it might be worth presenting here.

Lemma 7.1. *If for a weighted graph (Γ, μ) the conditions $(p_0), (VD), (H)$ hold, then*

$$E\left(x, R\right) \geq cR^2. \tag{7.7}$$

Proof   As we have seen in (7.6),

$$E\left(x, R\right) \geq c\rho\left(x, R/2, R\right) v\left(x, R/2, R\right)$$

follows from the conditions, and from (3.11) we obtain the statement.   ∎

7.2 Harnack graphs

The term of Harnack graphs was coined by Barlow (personal communication) some time ago in order to have a concise name for graphs satisfying the elliptic Harnack inequality. At that time, investigations were focused on fractals and fractal-like graphs in which the space-time scaling function was R^β. In this section we focus on graphs which on the one hand satisfy the elliptic Harnack inequality, and on the other hand satisfy the triplet $(p_0), (VD)$ already used in Section 7.1.

Proposition 7.3. *Assume that for (Γ, μ) $(p_0), (VD), (H)$ and (ρv) hold, then there is an $A = A_{\rho v} > 1$ such that anti-doubling $(aD\rho v)$ for ρv holds:*

$$\rho\left(x, AR, 2AR\right) v\left(x, AR, 2AR\right) \geq 2\rho\left(x, R, 2R\right) v\left(x, R, 2R\right) \tag{7.8}$$

for all $x \in \Gamma$.

Proof   Assume that $R > R_0$, otherwise the statement follows from (p_0). Let $A = B\left(x, R\right), B = B\left(x, 2R\right), D = B\backslash A$ where $R = 4kr$ for an $r \geq 1$. Let ξ_i denote the location of the first hit in $\partial B(x, (2\left(k + i\right)) 2r)$ for $i = 0...k - 1$. First, by Corollary 3.6

$$w_x\left(R\right) = \rho\left(x, R, 2R\right) v(x, R, 2R) \geq E_a\left(T_B\right).$$

It is evident that the exit time T_B in Γ^a satisfies

$$T_B \geq \sum_{i=0}^{k-1} T_{B(\xi_i, 2r)},$$

and consequently

$$E_a\left(T_B\right) \geq \sum_{i=0}^{k-1} E\left(\xi_i, 2r\right).$$

The terms on the right hand side can be estimated by using (H), as in (7.6), to obtain

$$E_a\left(T_B\right) \geq \sum_{i=0}^{k-1} E\left(\xi_i, 2r\right)$$

$$\geq \sum_{i=0}^{k-1} c \min_{z \in B} w_z\left(r\right) \geq ckw_x\left(r\right),$$

where (ρv) was used in the last step. Finally, choosing $A_{\rho v} = k$ so that $k \geq 2/c$, we get the statement. ∎

Remark 7.1. Let us observe that under the conditions of Proposition 7.3, with some increase of the number of iterations it follows that for the function

$$F\left(R\right) = \inf_{x \in \Gamma} \rho\left(x, R, 2R\right) v\left(x, R, 2R\right),$$

the anti-doubling property

$$F\left(A_F R\right) \geq 2F\left(R\right)$$

holds. Of course, the same applies for $F \simeq E$ in the presence of (E) because of Theorem 7.2.

Definition 7.2. *The resistance lower estimate* **RLE** (F) *holds for a function F if there is a $c > 0$ such that for all $x \in \Gamma, R > 0$*

$$\rho\left(x, R, 2R\right) \geq c \frac{F\left(x, 2R\right)}{V\left(x, 2R\right)}. \tag{7.9}$$

Proposition 7.4. *If for a weighted graph $\left(\Gamma, \mu\right)$ the conditions $\left(p_0\right), \left(VD\right), \left(H\right)$ and $RLE\left(E\right)$ hold, then $\left(ER\right)$ holds as well.*

Proof The lower estimate of E follows as above, the upper estimate is just $RLE\left(E\right)$. ∎

Proposition 7.5. *If for a weighted graph (Γ, μ) the conditions (p_0), (VD), (H) and $(aD\rho v)$ hold, then*

$$E(x, 2R) \simeq \rho(x, R, 2R) v(x, R, 2R)$$

Proof   We assume that $R > R_0$, otherwise the statement is a consequence of (p_0). The lower estimate can be deduced as in (7.6). The proof of the upper estimate uses Proposition 3.8. Let $M \geq A_{\rho v}, L = M^2$ be fixed constants ($A_{\rho v}$ is from Proposition 3.8), $R_k = M^k$, $B_k = B(x, R_k)$ and let n be the minimal integer so that $LR < R_n$. We have

$$E(x, 2R) \leq E(x, R_n) = \sum_{y \in B_n} g^{B_n}(x, y)\mu(y) \tag{7.10}$$

$$= \sum_{y \in B_0} g^{B_n}(x, y)\mu(y) + \sum_{m=0}^{n-1} \sum_{y \in B_{m+1} \setminus B_m} g^{B_n}(x, y)\mu(y). \tag{7.11}$$

As it follows from (p_0), the first term on the right hand side of (7.11) – the sum over B_0 – is majorized by a multiple of a similar sum over $B_1 \setminus B_0$ which is a part of the second term. Estimating g^{B_n} by (3.41), we obtain

$$E(x, 2R)$$
$$\leq E(x, LR)$$
$$\leq C \sum_{m=0}^{n} \left[\sum_{k=m}^{n} \rho(x, R_k, R_{k+1}) \right] v(x, R_m, R_{m+1})$$
$$\leq C \sum_{k=0}^{n} \left[\sum_{m=0}^{k} v(x, R_m, R_{m+1}) \right] \rho(x, R_k, R_{k+1})$$
$$\leq C \sum_{k=0}^{n} \rho(x, R_k, R_{k+1}) V(x, R_{k+1})$$
$$\leq C \sum_{k=0}^{n} \rho(x, R_k, R_{k+1}) v(x, R_k, R_{k+1}).$$

Now we use the anti-doubling property of ρv, which yields

$$\leq C\rho(R_{n-1}, R_n)v(R_{n-1}, R_n) \sum_{k=0}^{n} 2^{k-n}$$
$$\leq C\rho(R_{n-2}, R_{n-1})v(R_{n-2}, R_{n-1})$$
$$\leq C\rho(x, R, LR) v(x, R, LR),$$

where the last step uses the doubling properties of volume and resistance. ∎

Theorem 7.3. *If for a weighted graph (Γ, μ) the conditions (p_0), (VD), (H) and (ρv) hold, then*

$$E(x, 2R) \simeq \rho(x, R, 2R) v(x, R, 2R)$$

Proof   The statement is a direct consequence of Proposition 7.3 and 7.5. ∎

Remark 7.2. We can see that under $(p_0),(VD)$ and (H)

$$(wTC) \Longleftrightarrow (ER) \Longleftrightarrow (TC) \Longleftrightarrow (aD\rho v) \Longleftrightarrow RLE\,(E)\,.$$

The main line of the proof is indicated in the following diagrams:

$$(aD\rho v) + (VD) + (H) \Longrightarrow (ER),(TD),(TC),(wTC),(\overline{E})$$

which follows from Proposition (7.5), and

$$(wTC) + (VD) + (H) \Longrightarrow (ER),(TD),(TC),(\overline{E}),(aD\rho v)$$

which follows from Theorem (7.2).

Definition 7.3. *We introduce an upper and a lower bound for the Green kernel with respect to a function F. There are $C,c > 0$ such that for all $x \in \Gamma, R > 0, B = B\,(x,2R), A = B\,(x,R)\setminus B\,(x,R/2)$*

$$\max_{y\in A} g^B\,(x,y) \le C\frac{F\,(x,2R)}{V\,(x,R)}, \tag{7.12}$$

$$\min_{y\in A} g^B\,(x,y) \ge c\frac{F\,(x,2R)}{V\,(x,R)}. \tag{7.13}$$

If both are satisfied, it will be referred to as $g\,(F)$.

Theorem 7.4. *Assume that for a weighted graph (Γ,μ) the conditions $(p_0),(VD)$ hold. Then*

$$g\,(E) \Leftrightarrow (H) + (ER)$$

Proof   It is clear that $(7.12) + (7.13) \Longrightarrow HG\,(2)$ which is equivalent to (H) by Theorem 3.3. We know from Proposition 3.4 that $HG \Longrightarrow (H)$, and from Proposition 3.7 that (HG) implies (3.39). We can use $(3.39) + (7.13)$ to obtain

$$E\,(x,2R) \le C\rho\,(x,R,2R)\,v\,(x,R,2R)\,,$$

while the lower estimate follows from $(3.39) + (7.12)$, so we have (ER). The reverse implication follows from the fact that $(H) \Longrightarrow HG\,(2)$ which can be combined with (ER) to receive $g\,(E)$. ∎

Remark 7.3. One should note that (7.12) follows easily from the elliptic mean value inequality but (7.13) is stronger than a reversed kind of mean value inequality.

7.3 Strong anti-doubling property

The anti-doubling property has a stronger form, too.

Definition 7.4. *We say that the strong anti-doubling property holds for F if there are $B_F > A_F > 1$ such that for all $R \geq 1$*

$$F(A_F R) \geq B_F F(R). \tag{7.14}$$

This has the following equivalent form

$$\frac{F(x, R)}{F(x, r)} \geq c \left(\frac{R}{r}\right)^{\beta_1} \tag{7.15}$$

for some $c > 0, \beta_1 > 1$ and for all $x \in \Gamma, R > r > 0$.

In this section we deduce that (7.14) holds for $F(x, R) = \rho(x, R, 2R) v(x, R, 2R)$ working under the assumptions $(p_0), (VD)$ and (H). We will show that (7.14) or (7.15) for ρv follows if we assume that the graph is homogeneous with respect to the function ρv in $x \in \Gamma$.

Definition 7.5. *We define a new set of scaling functions V_a. $F \in V_a, a \in \{0, 1\}$ if there are $\beta' > a, c_F > 0$ such that for all $R > r > 0, x \in \Gamma, y \in B(x, R)$,*

$$\frac{F(x, R)}{F(y, r)} \geq c_F \left(\frac{R}{r}\right)^{\beta'}. \tag{7.16}$$

Lemma 7.2. *If (ER) holds then the following anti-doubling properties are equivalent (with different constants).*
1. There is an $A > 1$ such that

$$E(x, AR) \geq 2E(x, R) \tag{7.17}$$

for all x, R.
2. There is an $A' > 1$ such that

$$\rho(x, A'R, 2A'R)v(x, A'R, 2A'R) \geq 2\rho(x, R, 2R)v(x, R, 2R) \tag{7.18}$$

for all x, R.

Proof   Let us apply (ER) and (7.17) iteratively. Set $A' = A^k$ for some $k > 1$,

$$\rho(x, A'R, 2A'R)v(x, A'R, 2A'R)$$
$$\geq cE(x, A'R)$$
$$\geq c2^k E(x, R)$$
$$\geq c2^k c' \rho(x, R, 2R)v(x, R, 2R).$$

So if $k = \lceil -\log(cc') \rceil$, $A' = 2^k$, we receive (7.18). The reverse implication works in the same way.   ∎

For the strong anti-doubling property of $F(R) = \inf_{x \in \Gamma} E(x, R)$, first we show that it is at least linear.

Lemma 7.3. *If* (E) *holds, then for all* $L \in \mathbb{N}, R > 1$

$$F(LR) \geq LF(R).$$

Proof   Let us fix an $x = x_{\varepsilon, R}$ for which

$$F(LR) + \varepsilon \geq E(x, LR)$$

and let us use the strong Markov property:

$$E(x, LR) \geq E(x, (L-1)R) + \min_{z \in \partial B(x, (L-1)R)} E(z, R)$$

$$\geq \dots \geq L \min_{z \in B(x, LR)} E(x, R) \geq LF(R).$$

Since ε was arbitrary, we get the statement.   ■

Proposition 7.6. *If* $(p_0), (VD), (E)$ *and* (H) *hold, then there are* $B_F > A_F > 1$ *such that for all* $R \geq 1$,

$$F(A_F R) \geq B_F F(R). \tag{7.19}$$

Proof   The proof starts with choosing the reference point. Let $\varepsilon > 0$ be a small constant which will be chosen later. Assume that $R \geq 1$ and assign an $x = x_{\varepsilon, R} \in \Gamma$ to ε and R satisfying

$$F(3R) + \varepsilon \geq E(x, 3R).$$

Let us denote the first hitting time of a set $A = B(x, R)$ by τ_A and write $B = B(x, 3R), D = B(s, 2R)$. Also let $\xi = X_{T_D} \in \partial B(x, 2R)$ and let us split the history of the walk according to T_D. By using the strong Markov property, $E(x, 3R)$ can be estimated as follows

$$E(x, 3R) \geq E(x, 2R) + \mathbb{E}_x \left(\mathbb{E}_\xi \left[T_B \wedge \tau_A \right] \right)$$

$$+ \mathbb{E}(I \left[T_B > \tau_A \right] (T_B - \tau_A))$$

$$\geq F(2R) + \mathbb{E}_x \left(\mathbb{E}(\xi, R) \right)$$

$$+ \mathbb{E}_x \left[I(T_B > \tau_A) \mathbb{E}_\xi (T_B) \right]$$

$$\geq 2F(R) + F(R)$$

$$+ \mathbb{E}_x \left[I(T_B > \tau_A) \mathbb{E}_\xi (T_B) \right].$$

In the last step Lemma 7.3 was used. The third term contains the sub-case when the walk reaches $\partial B(x, 2R)$, then returns to A before leaving B. Let us denote this return site by $\zeta = X_k : k = \arg\min \{i > T_D, X_i \in A\}$. By using this, we get

$$\mathbb{E}_x \left[I \left(T_B > \tau_A \right) \mathbb{E}_\xi \left(T_B \right) \right]$$
$$= \mathbb{E}_x \left(\mathbb{P}_\xi \left(T_B > \tau_A \right) \right) E(\zeta, 2R))$$
$$\geq \mathbb{E}_x \left(P_\xi(T_B > \tau_A) F(2R) \right) \geq \min_{w \in \partial B(x, 2R)} \mathbb{P}_w(T_B > \tau_A) F(2R).$$

The probability in the above expression can be estimated by using the elliptic Harnack inequality (as in Theorem 3.4) to get

$$\min_{w \in \partial B(x, 2R)} \mathbb{P}_w(T_B > \tau_A) \geq c \frac{\rho(2R, 3R)}{\rho(R, 3R)} \geq c =: c_0.$$

We have come to the inequality

$$F(3R) + \varepsilon \geq 3F(R) + c_0 F(2R)$$
$$\geq 3F(R) + c_0 2F(R),$$

which means that if $c_F > \frac{2\varepsilon}{F(R)}$, i.e., $\varepsilon \leq \frac{F(1)c_0}{2}$, the statement follows with $A_F = 3$, $B_F = 3 + \frac{c_0}{2}$. ∎

Remark 7.4. Of course, we can formulate the strong anti-doubling property for $E(x, R)$ or ρv with a slight increase of A, but it seems more natural to state it for F.

Remark 7.5. It is also clear that F inherits $F(R) \geq cR^2$ from E or ρv.

7.4 Local space-time scaling

Now we introduce the set of functions that are candidates to be space-time scaling functions. We impose restrictions the space-time scaling functions which are as minimal as possible but strong enough to ensure the key properties of the mean exit time.

Definition 7.6. *We define W_0, a class of functions. $F \in W_0$ if $F : \Gamma \times \mathbb{R} \to \mathbb{R}$ and*
1. there are $\beta > 1, \beta' \geq 0, c_F, C_F > 0$ such that for all $R > r > 0, x \in \Gamma, y \in B(x, R)$

$$c_F \left(\frac{R}{r} \right)^{\beta'} \leq \frac{F(x, R)}{F(y, r)} \leq C_F \left(\frac{R}{r} \right)^\beta, \tag{7.20}$$

2. there is a $c > 0$ such that for all $x \in \Gamma, R > 0$

$$F(x, R) \geq cR^2, \tag{7.21}$$

3.

$$F(x, R + 1) \geq F(x, R) + 1 \tag{7.22}$$

for all $R \in \mathbb{N}$.

Finally, $F \in W_1$ if $F \in W_0$, and $\beta' > 1$ holds as well.

The function classes $W_1 \subset W_0$ will play a particular role in the whole sequel.

If $x = y$, we will refer to the upper (and lower) estimates in (7.20) as doubling (and anti-doubling) properties and if x, y are arbitrary, we will refer to the pair as doubling or regularity properties.

Clearly, it follows from (7.22) that $F : \Gamma \times \mathbb{N} \to \mathbb{R}$ has an inverse in the second variable $f(x, n) : \Gamma \times \mathbb{N} \to \mathbb{R}$:

$$f(x, n) = \min \{R \in \mathbb{N} : F(x, R) \geq n\} .$$

Similarly to Lemma 3.3, the relation (7.20) is equivalent to the followings. There are $C, c > 0, \beta \geq \beta' > 0$ such that for all $x \in \Gamma, n \geq m > 0, y \in B(x, e(x, n))$

$$c \left(\frac{n}{m}\right)^{1/\beta} \leq \frac{f(x, n)}{f(y, m)} \leq C \left(\frac{n}{m}\right)^{1/\beta'} . \tag{7.23}$$

Proposition 7.7. *For any (Γ, μ) assuming (p_0)*

$$(VD) + (TC) \Longrightarrow E \in W_0$$

Proof   It is clear that from (TC) follows the right hand side of (7.20). From Proposition 7.1 follows the left hand side of (7.20), the Einstein relation follows by Theorem 7.1, and its combination with the general inequality (3.11) implies (7.21). Monotonicity (7.22) is provided by (3.14)   ∎

8

Upper estimates

8.1 Some further heuristics

In Section 6 we have shown the local diagonal lower estimate. Now we would like to pave the way to a similar upper bound. We start with a well known, elementary argument. By definition,

$$\sum_{y \in \Gamma} p_n(x, y) \mu(y) = 1,$$

and by restricting the sum for a given ball $B(x, R)$, we have

$$\sum_{y \in B(x, R)} p_n(x, y) \mu(y) \leq 1,$$

$$\frac{1}{V(x, R)} \sum_{y \in B(x, R)} p_n(x, y) \mu(y) \leq \frac{1}{V(x, R)}.$$

If we know that the left hand side dominates the diagonal term which means that some kind of mean value inequality holds for it, we are ready if $R \simeq e(x, n)$:

$$cp_n(x, x) \leq \frac{1}{V(x, R)} \sum_{y \in B(x, R)} p_n(x, y) \mu(y) \leq \frac{1}{V(x, R)}.$$

So we need

$$p_n(x, x) \leq \frac{C}{V(x, R)} \sum_{y \in B(x, R)} p_n(x, y) \mu(y). \tag{8.1}$$

This type of mean value inequality is well known in classical diffusion theory and it is used for random walks as well (cf. [27]).

8.2 Mean value inequalities

Definition 8.1. *The space-time measure ν is induced by μ:*

$$\nu\left([s,t] \times A\right) = (t-s)\,\mu\left(A\right)$$

for $A \subset \Gamma, t \geq s \geq 0$.

Definition 8.2. *We say that* $\mathbf{PMV}_\delta\left(F\right)$, *(the strong form of) the parabolic mean-value inequality with respect to a function F holds on (Γ, μ), if for fixed constants $0 \leq c_1 < c_2 < c_3 < c_4 \leq c_5, 0 < \delta \leq 1$, there is a $C \geq 1$. such that for arbitrary $x \in \Gamma$, $n \in \mathbb{N}$ and $R > 0$, using the notations $F = F(x, R)$, $B = B(x, R)$, $\mathcal{D} = [0, c_5 F] \times B$, $\mathcal{D}^- = [c_1 F, c_2 F] \times B(x, \delta R)$, $\mathcal{D}^+ = [c_3 F, c_4 F] \times B(x, \delta R)$ for any non-negative Dirichlet sub-solution of the heat equation*

$$\triangle^B u \geq \partial_n u_n$$

on $\mathcal{D}$, the inequality

$$\max_{\mathcal{D}^+} u \leq \frac{C}{\nu\left(\mathcal{D}^-\right)} \sum_{(i,y)\in\mathcal{D}^-} u_i(y)\mu(y) \tag{8.2}$$

holds. (See Figure 8.1.)

Definition 8.3. *We say that* $\mathbf{wPMV}_\delta\left(F\right)$, *the weak parabolic mean-value inequality with respect to a function F holds on (Γ, μ), if for fixed constants $0 \leq c_1 < c_2 < c_3 \leq c_5, 0 < \delta \leq 1$, there is a $C \geq 1$. such that for arbitrary $x \in \Gamma$, $n \in \mathbb{N}$ and $R > 0$, using the notations $F = F(x, R)$, $B = B(x, R)$, $n = c_3 F$ $\mathcal{D} = [0, c_5 F] \times B$, $\mathcal{D}^- = [c_1 F, c_2 F] \times B(x, \delta R)$, for any non-negative Dirichlet sub-solution of the heat equation*

$$\triangle^B u \geq \partial_n u_n$$

on $\mathcal{D}$, the inequality

$$u_n\left(x\right) \leq \frac{C}{\nu\left(\mathcal{D}^-\right)} \sum_{(i,y)\in\mathcal{D}^-} u_i(y)\mu(y) \tag{8.3}$$

holds.

Definition 8.4. *We shall use $PMV(F)$ if $PMV_\delta(F)$ holds for $\delta = 1$.*

We shall omit the function F from the notations $PMV(F)$ if $F = E$, or from other circumstances it is clear that F is unique and there is no danger of confusion.

Definition 8.5. *The mean value inequality* $(\mathbf{MV})$ *holds if for all $x \in \Gamma, R > 0$ and $u \geq 0$ on $\overline{B}(x, R)$ is a harmonic function on $B = B(x, R)$,*

$$u\left(x\right) \leq \frac{C}{V\left(x, R\right)} \sum_{y \in B} u\left(y\right) \mu\left(y\right). \tag{8.4}$$

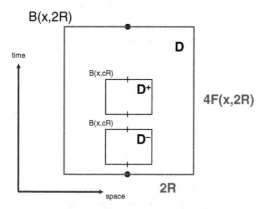

Fig. 8.1. The test balls for PMV

Definition 8.6. *We say that the mean-value property* (**MVG**) *holds for Green kernels on* (Γ, μ) *if there is a* $C \geq 1$. *such that for all* $R > 0, x \in \Gamma,$ $B = B(x, R)$ *and* $y \in \Gamma, d = d(x, y) > 0,$

$$g^B(y, x) \leq \frac{C}{V(x, d)} \sum_{z \in B(x, d)} g^B(y, z)\mu(z).$$ (8.5)

Definition 8.7. *We say that a Green kernel satisfies the upper estimate* $(\mathbf{g_{0,1}})$ *on* (Γ, μ) *if there is a* $C \geq 1$. *such that for all* $R > 0$ *and* $x, y \in \Gamma, d = d(x, y) > 0, B = B(x, R),$

$$g^B(y, x) \leq C\frac{E(x, R)}{V(x, d)}.$$ (8.6)

8.3 Diagonal estimates for strongly recurrent graphs

In this section we prove local diagonal estimates for recurrent graphs. The proof uses λ-resolvent which will be further extended to cover the transient case. Since the recurrent case is much simpler, it is useful to study it prior to the general case.

Theorem 8.1. *Assume that the random walk is* **strongly recurrent** *on* (Γ, μ)*: there is a* $c > 0$ *such that for all* $x \in \Gamma, R > 0,$

$$\rho(x, R, 2R) \geq c\rho(\{x\}, B^c(x, 2R)).$$ (8.1)

If $(p_0), (VD), (TC)$ *and* (MV) *holds, then there is a* $C > 0$ *such that for all* $x \in \Gamma, n > 0,$

$$P_n(x,x) \le C \frac{\mu(x)}{V(x,e(x,n))}.$$

Let us define the λ-resolvent as follows:

$$G_\lambda(x,x) = \sum_{k=0}^{\infty} e^{-\lambda k} P_k(x,x).$$

The starting point of the proof of the DUE is the following lemma for the $\lambda-$resolvent.

Lemma 8.1. *In general if $\lambda^{-1} = n$, then*

$$P_{2n}(x,x) \le c\lambda G_\lambda(x,x).$$

Proof   The proof is elementary. From eigenfunction decomposition it follows that $P_{2n}^{B(x,R)}(x,x)$ is non-increasing in n. For $R > 2n$, $P_{2n}^{B(x,R)}(x,x) = P_{2n}(x,x)$, hence the monotonicity holds for $P_{2n}(x,x)$ in the $2n < R$ time range. But R is chosen arbitrarily, hence P_{2n} is non-increasing and we derive

$$G_\lambda(x,x) = \sum_{k=0}^{\infty} e^{-\lambda k} P_k(x,x) \ge \sum_{k=0}^{\infty} e^{-\lambda 2k} P_{2k}(x,x) \ge \sum_{k=0}^{n-1} e^{-\lambda 2k} P_{2k}(x,x)$$

$$\ge P_{2n}(x,x) \frac{1 - e^{-\lambda 2n}}{1 - e^{-2\lambda}}.$$

Choosing $\lambda^{-1} = n$ the statement follows.   ∎

Lemma 8.2. *If $(\overline{E})$ holds and $1/\lambda \le \frac{1}{2}E(x,R)$, then*

$$G_\lambda(x,x) \le cG^B(x,x),$$

where $B = B(x,R)$

Proof   Let ξ_λ be a geometrically distributed random variable with parameter $e^{-\lambda}$. We can easily see that

$$G^B(x,x) = G_\lambda(x,x) + E_x(I(T_{x,R} \ge \xi_\lambda)G^B(X_{\xi_\lambda},x))$$
$$-E_x(I(T_{x,R} < \xi_\lambda)G_\lambda(X_{T_{x,R}},x)), \tag{8.2}$$

and

$$G_\lambda(x,x) \le P(T_{x,R} \ge \xi_\lambda)^{-1}G^B(x,x).$$

Here $P(T_{x,R} \ge \xi_\lambda)$ can be estimated thanks to (6.13),

$$P(T_{x,R} \ge \xi_\lambda) \ge P(T_{x,R} > n, \xi_\lambda \le n)$$

$$\ge P(\xi_\lambda \le n)P(T_{x,R} > n) \ge c' \frac{E-n}{2C\overline{E}} > c$$

if $\lambda^{-1} \le n = \frac{1}{2}E(x,R)$ and $(\overline{E})$ holds.   ∎

Proof [of Theorem 8.1] Combining the previous lemmas for $\lambda^{-1} = n = \frac{1}{2}E(x, R), B = B(x, R)$ we obtain

$$P_{2n}(x, x) \leq c\lambda G_\lambda(x, x) \leq cE(x, R)^{-1}G^R(x, x).$$

Now let us recall from (3.4) that $G^B(x, x) = \mu(x)\rho(\{x\}, B^c)$ and let us use the conditions to obtain

$$G^B(x, x) = \mu(x)\rho(\{x\}, B^c) \leq C\mu(x)\rho(x, R, 2R)$$

$$\leq \frac{C\mu(x)}{\lambda(x, R)V(x, R)} \leq \frac{C\mu(x)\overline{E}(x, R)}{V(x, R)},$$

and from Lemmas 8.1 and 8.2 it follows that

$$P_{2n}(x, x) \leq C\mu(x)E(x, R)^{-1}\rho(\{x\}, B^c)$$

$$\leq \frac{C\mu(x)\overline{E}(x, R)}{E(x, R)V(x, R)} \leq \frac{C\mu(x)}{V(x, R)} \leq \frac{C\mu(x)}{V(x, e(x, n))}.$$

$\blacksquare$

8.4 Local upper estimates and mean value inequalities

After all these preparations, we can state the main result of this chapter.

Theorem 8.2. *Assume that (Γ, μ) satisfies $(p_0), (VD)$ and (TC), then the following statements are equivalent.*

1. *The local diagonal upper estimate $DUE(E)$ holds, i.e.*

$$p_n(x, x) \leq \frac{C}{V(x, e(x, n))}; \tag{8.3}$$

2. *the local particular upper estimate $\mathbf{PUE}(E)$ holds, i.e.,*

$$p_n(x, y) \leq C\left(\frac{1}{V(x, e(x, n))V(y, e(y, n))}\right)^{\frac{1}{2}}; \tag{8.4}$$

3. *the upper estimate $\mathbf{UE}(E)$ holds, i.e.*

$$p_n(x, y) \leq C\frac{\exp\left[-c\left(\frac{E(x, d(x, y))}{n}\right)^{\frac{1}{\beta-1}}\right]}{V(x, e(x, n))}; \tag{8.5}$$

4. *$PMV(E)$ holds;*
5. *$wPMV(E)$ holds;*
6. *(MV) holds;*
7. *(MVG) holds;*
8. *(g_{01}) holds.*

The most sophisticated part of the proof of Theorem 8.2 is showing $(g_{01}) \implies DUE(E)$ and $UE(E)$. This will be given in Section 8.5. The equivalence of the other conditions will be shown in Section 8.8.

8.5 λ, m-resolvent

In this section we present a refined technique of the resolvent method which was developed in [7] and [49] to provide a unified treatment for recurrent and transient diagonal upper estimates.

8.5.1 Definition of λ, m-resolvent

For any non-empty finite set $B \subset \Gamma$, for all real $\lambda \geq 0$ and integer $m \geq 0$, define the λ, m-resolvent of Δ in B as

$$G^B_{\lambda,m} = \left(\lambda I - \Delta^B\right)^{-m}. \tag{8.6}$$

It follows from the definition that

$$G^B_{\lambda,0} = I \quad \text{and} \quad G^B_{0,1} = G^B,$$

where G^B is the usual Green function in B. Clearly, for any $m \geq 1$, we have

$$G^B_{\lambda,1} \circ G^B_{\lambda,m-1} = G^B_{\lambda,m}. \tag{8.7}$$

Since $\Delta^B = P^B - I$, from (8.6) we obtain

$$G^B_{\lambda,m} = \left((\lambda+1)I - P^B\right)^{-m} = \omega^m \left(I - \omega P^B\right)^{-m},$$

where $\omega = (\lambda+1)^{-1}$. Expanding it by Taylor's formula, we obtain

$$G^B_{\lambda,m} = \omega^m \left(I + m\omega P^B + \frac{m(m+1)}{2}\omega^2 \left(P^B\right)^2 + \ldots\right) = \sum_{n=0}^{\infty} Q_m(n)\omega^{n+m} P^B_n, \tag{8.8}$$

where

$$Q_m(n) = \begin{cases} 1 & m = 0, n = 0, \\ 0 & m = 0, n \geq 1, \\ \binom{n+m-1}{m-1}, & m \geq 1. \end{cases} \tag{8.9}$$

By the formula (8.8) we extend the definition of $G^B_{\lambda,m}$ to infinite sets B. If $\lambda > 0$, then the series in (8.8) always converge and $G^B_{\lambda,m} < \infty$. If $\lambda = 0$, then $G^B_{\lambda,m}$ may be equal to ∞.

The λ, m-resolvent obviously has a symmetric kernel defined as

$$g^B_{\lambda,m}(x,y) = \frac{G^B_{\lambda,m}(x,y)}{\mu(y)} = \sum_{n=0}^{\infty} Q_m(n)\omega^{n+m} p^B_n(x,y).$$

If B is finite and $m \geq 1$, then (8.6) implies

$$(\Delta - \lambda) g^B_{\lambda,m} = -g^B_{\lambda,m-1} \quad \text{in } B. \tag{8.10}$$

It is easy to see that by considering $G^B_{\lambda,m}$ as an operator, we have

$$\left\|G^B_{\lambda,m}\right\| \leq \left\|G^B_{0,m}\right\| \leq \overline{E}(x,R)^m. \tag{8.11}$$

8.5.2 Upper bound for the $0, m$-resolvent

Lemma 8.3. *Let the graph (Γ, μ) satisfy the conditions $(VD) + (g_{0,1})$. Then for any $m \geq 1$, for all $x \in \Gamma$, $R > 0$ we have*

$$g_{0,m}^{B(x,R)}(x,y) \leq C \frac{\overline{E}(x,5R)^m}{V(x,d)}. \tag{8.12}$$

Proof   The case $m = 1$ follows from $(g_{0,1})$, so we can assume $m \geq 2$ and argue by induction in m. For simplicity write $B = B(x,R)$, $G_m^B = G_{0,m}^B$ and $g_m^B = g_{0,m}^B$. Assuming $y \in B(x,R)$, let us set $r = d(x,y)/2$ and observe that the balls $B(x,r)$ and $B(y,r)$ do not intersect. Therefore, by (8.7),

$$g_m^B(x,y) = \sum_z g_{m-1}^B(x,z) g_1^B(z,y)\mu(z)$$

$$\leq \left(\sum_{z \notin B(x,r)} + \sum_{z \notin B(y,r)} \right) g_{m-1}^B(x,z) g_1^B(z,y)\mu(z).$$

Writing

$$f(z) = g_1^B(z,y)\mathbf{1}_{\{z \notin B(y,r)\}} \quad \text{and} \quad h(z) = g_{m-1}^B(x,z)\mathbf{1}_{\{z \notin B(x,r)\}}$$

we obtain

$$g_m^B(x,y) \leq G_{m-1}^B f(x) + G_1^B h(y). \tag{8.13}$$

Since the Green kernels in question are symmetric and increase with B, by $(g_{0,1})$ and (VD) we obtain,

$$\|f\| = \sup_{z \notin B(y,r)} g_1^{B(x,R)}(z,y) \leq \sup_{z \notin B(y,r)} g_1^{B(y,2R)}(y,z) \leq C \frac{E(y,4R)}{V(y,r)} \leq C \frac{\overline{E}(x,5R)}{V(x,r)},$$

and by the inductive hypothesis we get

$$\|h\| = \sup_{z \notin B(x,r)} g_{m-1}^{B(x,R)}(x,z) \leq C \frac{\overline{E}(x,5R)^{m-1}}{V(x,r)}.$$

By combining (8.13), (8.11) and the above estimates for $\|f\|, \|g\|$, we obtain (8.12).   ∎

8.5.3 Feynman-Kac formula for polyharmonic functions

Let B be a non-empty finite subset of Γ. Let f be a function on Γ such that

$$\Delta f - \lambda f = 0 \quad \text{in } B. \tag{8.14}$$

Then the function $v = G_{\lambda,m}^B f$ satisfies

$$(\varDelta - \lambda)^{m+1} v = 0 \quad \text{in } B. \tag{8.15}$$

Therefore, v_m is a λ-*polyharmonic function*. Moreover, v satisfies the following boundary conditions outside B:

$$v = 0, \ (\varDelta - \lambda) v = 0, \ (\varDelta - \lambda)^2 v = 0, \ ..., \ (\varDelta - \lambda)^m v = (-1)^m f, \quad (8.16)$$

so that v can be regarded as a solution to the boundary value problem (8.15)–(8.16). Conversely, any solution to this problem is given by

$$v = G^B_{\lambda, m} f,$$

where the boundary function f has to be extended first to B to satisfy (8.14). The following statement can be considered as a probabilistic representation of the solution to (8.15)–(8.16).

Lemma 8.4. (Feynman-Kac formula) *Let f be a function on Γ satisfying*

$$\varDelta f - \lambda f = 0 \quad \text{in } B. \tag{8.17}$$

Then for any $x \in B$,

$$f(x) = \mathbb{E}_x \left[\omega^T f(X_T) \right] \tag{8.18}$$

and furthermore, for any $m \geq 0$,

$$G^B_{\lambda, m} f(x) = \mathbb{E}_x \left[Q_{m+1}(T) \omega^{T+m} f(X_T) \right], \tag{8.19}$$

where $T = T_B$ is the first exit time from B.

Proof   For any integer $m \geq -1$, write

$$v_m(x) = \mathbb{E}_x \left[Q_{m+1}(T) \omega^{T+m} f(X_T) \right]. \tag{8.20}$$

Clearly, $v_{-1} = 0$ in B, $v_0 = f$ in B^c and $v_m = 0$ in B^c for all $m \geq 1$. Let us prove that for all $m \geq 0$,

$$\varDelta v_m - \lambda v_m = -v_{m-1} \quad \text{in } B. \tag{8.21}$$

Indeed, for any $x \in B$, the Markov property implies

$$\mathbb{E}_x \left[Q_{m+1}(T-1) \omega^{T-1+m} f(X_T) \right] = \tag{8.22}$$

$$\sum_{y \sim x} P(x, y) \mathbb{E}_y \left[Q_{m+1}(T) \omega^{T+m} f(X_T) \right] = P v_m(y). \tag{8.23}$$

By the property of binomial coefficients, for all $m \geq 0$ and $T \geq 1$ we have

$$Q_{m+1}(T-1) = Q_{m+1}(T) - Q_m(T).$$

Hence, the left hand side of (8.22) is equal to

$$\mathbb{E}_x \left[Q_{m+1}(T) \omega^{T-1+m} f(X_T) \right] - \mathbb{E}_x \left[Q_m(T) \omega^{T-1+m} f(X_T) \right]$$
$$= \omega^{-1} v_m(x) - v_{m-1}(x).$$

By substituting this in (8.22) and using $\omega^{-1} = 1 + \lambda$, we obtain (8.21).

For $m = 0$, we obtain $\Delta v_0 - \lambda v_0 = 0$ from (8.21). Since $v_0 = f$ outside B, we obtain $v_0 = f$ also in B. Therefore, (8.18) follows from (8.20) for $m = 0$. If $m \geq 1$, then by solving (8.21) with the boundary condition $v_m = 0$ outside B, we obtain $v_m = G_\lambda^B v_{m-1}$. Therefore, $v_m = G_{\lambda,m}^B f$, whence (8.19) follows. ∎

Corollary 8.1. *For any non-empty finite set $B \subset \Gamma$ and for any non-negative function f in Γ such that $\Delta f - \lambda f = 0$ in B,*

$$\left(G_{\lambda,m}^B f(x) \right)^2 \leq c_m f(x) G_{\lambda,2m}^B f(x) \tag{8.24}$$

for all $x \in \Gamma$ and for all $\lambda \geq 0$, $m \geq 0$.

Proof   Using notation (8.20), by the Cauchy-Schwarz inequality we have

$$v_m(x)^2 \leq \mathbb{E}_x \left[\omega^T f(X_T) \right] \mathbb{E}_x \left[Q_{m+1}^2(T) \omega^{T+2m} f(X_T) \right]. \tag{8.25}$$

By (8.9), we obtain $Q_{m+1}^2(T) \leq c_m Q_{2m+1}(T)$. Hence (8.25) implies $v_m^2 \leq c_m v_0 v_{2m}$ which, by Lemma 8.4, coincides with (8.24). ∎

Lemma 8.5. *Assume that $(\overline{E})$ holds on (Γ, μ). Let $B = B(x,r)$ and let v be a function on $\overline{B}$ such that $0 \leq v \leq 1$. Suppose that in B v satisfies the equation*

$$\Delta v = \lambda v, \tag{8.26}$$

where λ is a constant such that

$$\overline{E}(x, R)^{-1} \leq \lambda. \tag{8.27}$$

Then

$$v(x) \leq e^{-c^*}, \tag{8.28}$$

where $c^* = -\log(1 - \varepsilon)$, $\varepsilon = \frac{1}{2\overline{C}} > 0$, $\overline{C}$ is the constant in the condition $(\overline{E})$ (see Figure 8.2).

Lemma 8.6. *Assume that (Γ, μ) satisfies $(\overline{E})$. Let $B = B(x, R)$ be an arbitrary non-empty ball on Γ, and let v be a function on $\overline{B}$ such that $0 \leq v \leq 1$. If v in B satisfies the equation*

$$Pv = \lambda v$$

with a constant λ such that

$$\overline{E}(x, R)^{-1} \leq \lambda, \tag{8.29}$$

then

$$v(x) \leq \exp \left[-\frac{c^*}{2} \frac{R}{e(\underline{y}, \lambda^{-1})} \right]. \tag{8.30}$$

Here $\underline{y} \in B$ maximizes $e(y, \lambda^{-1})$ in $y \in B$ and $c^* > 0$ is from Lemma 8.5.

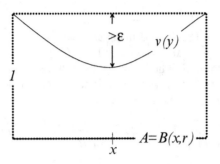

Fig. 8.2. The depression of v

Proof   From the assumption (8.29) and Lemma 8.5 we obtain $v(x) \leq e^{-c^*}$. If in addition we have

$$R \leq e\left(\underline{y}, \lambda^{-1}\right), \tag{8.31}$$

then (8.30) is trivially satisfied. In particular, if R is in the bounded range, then (8.31) is true because λ is bounded from above by (8.29). Hence in the sequel we may assume that

$$R > C \text{ and } R > e\left(\underline{y}, \lambda^{-1}\right), \tag{8.32}$$

or in another form

$$\lambda^{-1} < E\left(\underline{y}, R\right) \leq E\left(x, R\right).$$

Let us choose a number r as $r = e\left(\underline{y}, \lambda^{-1}\right)$. By the definition of $\underline{y}$ Lemma 8.5 applies in any ball $B\left(z, r\right), z \in B\left(x, R\right)$ of radius r since $e\left(z, \lambda^{-1}\right) \leq e\left(\underline{y}, \lambda^{-1}\right) = r$ which means

$$\lambda^{-1} \leq E\left(z, r\right) \leq \overline{E}\left(z, r\right).$$

Let x_i, $i \geq 1$ be a point in the ball $B(x, (r+1)i)$, where v takes the maximum value in this ball, and let $m_i = v(x_i)$. For $i = 0$, we set $m_0 = v(x)$. For each $i \geq 0$, consider the ball $A_i = B(x_i, r)$. Since

$$\overline{A}_i \subset B(x_i, r+1) \subset B(x, (r+1)(i+1)),$$

we have

$$\max_{\overline{A}_i} v \leq m_{i+1}.$$

By applying Lemma 8.5 to the function v/m_{i+1} in the ball A_i, we obtain

$$m_i \leq (1 - \varepsilon)m_{i+1}.$$

By iterating this inequality $k = \lfloor R/(r+1) \rfloor$ times and using $m_k \leq 1$, we arrive at

$$v(x) = m_0 \leq (1 - \varepsilon)^k = e^{-c^* k}. \tag{8.33}$$

Under the conditions (8.32) and (8.29), we have

$$k \geq \frac{R}{2r} = \frac{R}{2e\left(\underline{y}, \lambda^{-1}\right)},$$

hence (8.30) follows from (8.33).   ∎

Corollary 8.2. *Assume that (Γ, μ) satisfies $(\overline{E})$. Let f be a non-negative function on Γ which satisfies in the ball $B = B(x, R)$ the equation $\Delta f - \lambda f = 0$ with a constant $\lambda \in (0, 1)$. Then f admits the following estimate at the point x:*

$$f(x) \leq C \exp\left[-c \frac{R}{e\left(\underline{y}, \lambda^{-1}\right)}\right] \max_{B^c} f. \tag{8.34}$$

Here $C, c > 0$ are constants depending on the conditions.

Proof   Thanks to the maximum principle, $f(y) \leq \sup_{B^c} f$ for $y \in B$ and from applying Lemma 8.6 to $v = f / \sup_{B^c} f$ the statement follows. The restriction $\lambda^{-1} \leq \overline{E}(x, R)$ can be eliminated since if $\lambda^{-1} > \overline{E}(x, R)$, then by definition $e\left(\underline{y}, \lambda^{-1}\right) \geq e\left(x, \lambda^{-1}\right) > R$, and $\frac{R}{e(\underline{y}, \lambda^{-1})} < 1$ and C can be adjusted to satisfy (8.34).   ∎

Corollary 8.3. *Under the conditions of Lemma 8.6, we have*

$$G_{\lambda,l}^B f(x) \leq C E(x, R)^l \exp\left[-c \frac{R}{e\left(\underline{y}, \lambda^{-1}\right)}\right] \sup_{B^c} f,$$

for any non-negative integer l.

Proof   The statement follows from Corollary 8.1, 8.2 and (8.11).   ∎

8.5.4 Upper bound for λ, m-resolvent

Theorem 8.3. *Assuming (p_0) the conditions $(VD), (TC)$ and $(g_{0,1})$, it follows that for a large enough $m > 1$ and for all $0 < \lambda < 1$, $x \in \Gamma$, the inequality $(g_{\lambda,m})$*

$$g_{\lambda,m}(x, x) \leq C \frac{\lambda^{-m}}{V(x, e(x, \lambda^{-1}))} \tag{8.35}$$

holds.

We start with a lemma.

Lemma 8.7. *Assuming that (Γ, μ) satisfies (p_0), (VD), (TC) and $(g_{0,1})$, for all $x \in \Gamma$, $0 < r < R$, we have*

$$g_{\lambda,m}^{B(x,R)}(x,x) - g_{\lambda,m}^{B(x,r)}(x,x) \leq C \frac{E(x,R)^m}{V(x,r)} \exp\left[-c\frac{R}{e\left(\underline{y},\lambda^{-1}\right)}\right],$$

where $\underline{y}$ maximizes $e\left(y,\lambda^{-1}\right)$ in $y \in B(x,R)$

Proof   For simplicity, write $A = B(x,R)$, $B = B(x,r)$ and set

$$v_m(y) = g_{\lambda,m}^A(x,y) - g_{\lambda,m}^B(x,y).$$

Clearly, we have $v_0 = 0$ and $\Delta v_m - \lambda v_m = -v_{m-1}$ in B for $m \geq 1$. Therefore,

$$v_m = G_\lambda^B v_{m-1} + u_m \quad \text{in } B, \tag{8.36}$$

where u_m is defined by

$$\begin{cases} \Delta u_m - \lambda u_m = 0 \text{ in } B, \\ u_m|_{B^c} = v_m. \end{cases}$$

By iterating (8.36) and using $v_0 = 0$, we obtain

$$v_m = G_{\lambda,m-1}^B u_1 + G_{\lambda,m-2}^B u_2 + \ldots + u_m. \tag{8.37}$$

By Corollary 8.3, we have that

$$G_{\lambda,l}^B u_i(x) \leq C E(x,r)^l \exp\left[-c\frac{r}{e\left(\underline{y},\lambda^{-1}\right)}\right] \sup_{A \backslash B} v_i, \tag{8.38}$$

and Lemma 8.3 yields

$$\sup_{A \backslash B} v_i = \sup_{y \in A \backslash B} g_{\lambda,i}^A(x,y) \leq C \frac{E(x,R)^i}{V(x,r)}. \tag{8.39}$$

Therefore, from (8.37), (8.38), and (8.39) we obtain

$$v_m(x) = \sum_{i=1}^m G_{\lambda,m-i}^B u_i(x) \leq C \sum_{i=1}^m E(x,r)^{m-i} \exp\left[-c\frac{R}{e\left(\underline{y},\lambda^{-1}\right)}\right] \sup_{A \backslash B} v_i$$

$$\leq C \frac{E(x,R)^m}{V(x,r)} \exp\left[-c\frac{R}{e\left(\underline{y},\lambda^{-1}\right)}\right],$$

which was to be proved.   ∎

Proof [of Theorem 8.3] From the inequality (7.5) (which follows from the conditions by using Proposition 7.1), we have the constant A_E and without loss of generality we can assume that $A_E \geq 2$. Let $R_k = A_E^k$ and $B_k = B(x, A_E^k), E_k = E(x, R_k)$. We have

$$g_{\lambda, m}(x, x) = g_{\lambda, m}^{B_0}(x, x) + \sum_{k=0}^{\infty} \left(g_{\lambda, m}^{B_{k+1}}(x, x) - g_{\lambda, m}^{B_k}(x, x) \right).$$

The term $g_{\lambda, m}^{B_0}(x, x)$ can be estimated as follows:

$$g_{\lambda, m}^{B_0}(x, x) = \sum_{n=0}^{\infty} Q_m(n) \omega^{n+m} P_n^{B(x,1)}(x, x) = Q_m(0) \omega^m P_0^{B(x,1)}(x, x) = \omega^m \leq 1.$$

By Lemma 8.7 we obtain

$$g_{\lambda, m}(x, x) \leq 1 + C \sum_{k=0}^{\infty} \exp\left[-c \frac{R_k}{e\left(\underline{y}_k, \lambda^{-1}\right)} \right] \frac{E_{k+1}^m}{V_k}. \qquad (8.40)$$

where $\underline{y}_k$s maximize the term $e\left(y, \lambda^{-1}\right)$ in B_k. Let us observe that we have used $(g_{0,1})$ only for $y \in \Gamma, d(x, y) \geq \frac{1}{A_F} R$.

Let $R = e\left(x, \lambda^{-1}\right)$, $E = E(x, R)$ and let l be the smallest integer for which $\lambda E_l > 1$. For all $i < l$, we use

$$E_i \leq 2^{i-l} E_l$$

and

$$V_l \leq A_E^{(l-i)\alpha} V_i$$

to get

$$\sum_{i=0}^{l-1} \exp\left[-c \frac{R_i}{e\left(\underline{y}_i, \lambda^{-1}\right)} \right] \frac{E_i^m}{V_i} \leq C \sum_{i=0}^{l-1} \left(\frac{2^m}{A_E^\alpha} \right)^{i-l} \frac{E^m}{V(x, R)}$$

$$\leq C \frac{E^m}{V(x, R)} \sum_{i=0}^{l-1} \left(\frac{2^m}{A_E^\alpha} \right)^{i-l} \leq C \frac{E^m}{V(x, R)}$$

if $m > \alpha \log_2 A_E$. The upper part of the sum can be estimated as follows. Write $n = \lfloor \lambda^{-1} \rfloor$ and observe that from (7.5) (which follows from (wTC), hence from (TC) as well) it follows that

$$R_i \geq A_E^{i-l-1} R = A_E^{i-l-1} e(x, n) \geq ce\left(x, 2^{i-l} n\right),$$

while from (TC) we get

$$\frac{e\left(x, 2^{i-l} n\right)}{e\left(\underline{y}_i, \lambda^{-1}\right)} \geq c2^{\frac{i-l}{\beta}}$$

and

$$\sum_{i=l}^{\infty} \exp\left[-\frac{R_i}{e\left(\underline{y}_i, \lambda^{-1}\right)}\right]\frac{E_i^m}{V_i} \le C\sum_{i=l}^{\infty}\exp\left[-c2^{\frac{i-l}{\beta}}\right]\left(\frac{A_E^i}{R}\right)^{\beta m}\frac{V(x,R)}{V(x,A_E^i)}\frac{E^m}{V(x,R)}$$

(8.41)

$$\le C\frac{E^m}{V(x,R)}\sum_{i=l}^{\infty}\exp\left[-c2^{\frac{i-l}{\beta}}\right]\left(\frac{A_E^i}{R}\right)^{\beta m} \le C\frac{E^m}{V(x,R)}\sum_{i=0}^{\infty}\exp\left[-c2^{\frac{i}{\beta}}\right]\left(A_E^m\right)^{\beta i}.$$

(8.42)

It is clear that the sum we have obtained is bounded by a constant depending on A_E, β since the sum decreases faster than a geometric series. Finally, from (VD) it follows that $V(x,R) \le CR^\alpha \le CE(x,R)^m$ if $m > \alpha \log_2 A_E$ which means that the term 1 in (8.40) is also bounded by $C\frac{E(x,R)^m}{V(x,R)}$. Hence

$$g_{\lambda,m}(x,x) \le C\frac{E(x,R)^m}{V(x,R)} = C\frac{\lambda^{-m}}{V(x,e(x,\lambda^{-1}))},$$

which was to be proved.   ∎

Remark 8.1. In fact, we have used Lemma 8.7 only for $r = R/A_E$ which means that in all our conditions we can assume $d(x,y) \ge cR$.

8.6 Diagonal upper estimates

Now we are in the position that we can prove the diagonal upper estimate.

Theorem 8.4. *If (Γ, μ) satisfies (VD) and $(g_{\lambda,m})$, then for all $x, \in \Gamma, n \ge 1$*

$$p_n(x,x) \le \frac{C}{V(x,e(x,n))}$$

(8.43)

and

$$p_n(x,y) \le \frac{C}{\sqrt{V(x,e(x,n))V(y,e(y,n))}}$$

(8.44)

hold.

Lemma 8.8. *For all random walks on weighted graphs, $x,y \in A \subseteq \Gamma, n, m \ge 0$,*

$$p_{n+m}^A(x,y) \le \sqrt{p_{2n}^A(x,x)p_{2m}^A(y,y)}.$$

(8.45)

Proof   The proof is standard:

$$p_{n+m}^A(x,y) = P_{n+m}^A(x,y)\frac{1}{\mu(y)} = \sum_z P_n^A(x,z)\left(\frac{\mu(z)}{\mu(z)}\right)^{1/2} P_m^A(z,y)\frac{1}{\mu(y)}$$

$$\leq \left(\sum_z P_n^A(x,z)^2\frac{1}{\mu(z)}\right)^{1/2}\left(\sum_z \frac{\mu(z)}{\mu^2(y)}P_m^A(z,y)^2\right)^{1/2}$$

$$\leq \left(\sum_z P_n^A(x,z)P_n^A(z,x)\frac{1}{\mu(x)}\right)^{1/2}\left(\sum_z P_m^A(x,z)P_m(z,y)\frac{1}{\mu(y)}\right)^{1/2}$$

$$= \left(P_{2n}^A(x,x)\frac{1}{\mu(x)}\right)^{1/2}\left(P_{2m}^A(y,y)\frac{1}{\mu(y)}\right)^{1/2}.$$

∎

Proof   [of Theorem 8.4] Let us first prove the following inequality: if $\lambda = n^{-1}$, then

$$p_{2n}(x,x) \leq C\lambda^m g_{\lambda,m}(x,x). \tag{8.46}$$

Indeed, $p_{2k}(x,x)$ is non-increasing in k. Therefore, for $\lambda = n^{-1}$ we have

$$g_{\lambda,m}(x,x) = \sum_{k=0}^\infty Q_m(k)\omega^k p_k(x,x) \geq \sum_{k=0}^\infty Q_m(2k)\omega^{2k}p_{2k}(x,x)$$

$$\geq c\sum_{k=1}^n k^{m-1}\omega^{2k}p_{2k}(x,x) \geq cn^m e^{-2}p_{2n}(x,x),$$

(where we have used $\omega^{2k} \geq (1+1/n)^{-2n} \geq e^{-2}$) whence (8.46) follows. By the hypothesis $(g_{\lambda,m})$, we have

$$g_{\lambda,m}(x,x) \leq \frac{C}{\lambda^m V(x,e(x,\lambda^{-1}))},$$

which, together with (8.46) implies DUE for even n. To prove PUE for even n, apply Lemma 8.8:

$$p_{2n}(x,y) \leq \sqrt{p_{2n}(x,x)p_{2n}(y,y)} \leq \frac{C}{\sqrt{V(x,e(x,n))V(y,e(y,n))}},$$

and PUE follows. By the semigroup property, we have

$$p_{2n+1}(x,y) = \sum_{z\sim x} P_1(x,z)p_{2n}(z,y) \leq \sup_{z\sim x} p_{2n}(z,y), \tag{8.47}$$

whence by (p_0)

$$p_{2n+1}(x,y) \leq \sup_{z\sim x} p_{2n}(z,y) \leq \sup_{z\sim x}\frac{C}{\sqrt{V(z,e(x,n))V(y,e(y,n))}}.$$

By (VD) and (TC), $V(z,e(z,n)) \simeq V(x,e(x,n))$ for all $z \sim x$, therefore, $PUE(E)$ and $DUE(E)$ follow for odd n. Finally, $PUE(E)$ and $DUE(E)$ for $n = 1$ follow directly from the definition of $p_n(x,y)$.   ∎

8.7 From DUE to UE

The proof is an easy modification of the nice argument given in [45].

Lemma 8.9. *Let* $r = \frac{1}{2} d(x, y)$, *then*

$$
p_{2n}(x, y) \leq P_x(T_{x,r} < n) \max_{\substack{n \leq k \leq 2n \\ v \in \partial B(x,r)}} p_k(v, y) + P_y(T_{y,r} < n) \max_{\substack{n \leq k \leq 2n \\ z \in \partial B(y,r)}} p_k(z, x).
$$

$$(8.48)$$

Proof   The statement follows from the first exit decomposition starting from x (and from y respectively), and from the Markov property.   ∎

Exercise 8.1. The development of the proof is left to the reader as an exercise.

Theorem 8.5. $(p_0) + (VD) + DUE(E) \Longrightarrow UE(E)$.

Lemma 8.10. *If* $(p_0), (VD)$ *and* $F \in W_0$ *hold, then for all* $\varepsilon > 0$ *there are* $C_\varepsilon, C > 0$ *such that for all* $k > 0, y, z \in \Gamma, r = d(y, z)$

$$
\sqrt{\frac{V(y, f(y, k))}{V(v, f(z, k))}} \leq C_\varepsilon \exp\left[\varepsilon C \left(\frac{F(y, r)}{k}\right)^{\frac{1}{(\beta-1)}}\right].
$$

Proof   Let us consider the minimal m for which $f(y, m) \geq r$,

$$
f(y, k) \leq f(y, k + m)
$$

and let us use the anti-doubling property with $\beta' > 0$ to obtain

$$
\sqrt{\frac{V(y, f(y, k))}{V(z, f(z, k))}} \leq \sqrt{\frac{V(y, f(y, k+m))}{V(z, f(z, k))}}
$$

$$
\leq C\left(\frac{f(y, k+m)}{f(z, k)}\right)^{\alpha/2} \leq C\left(\frac{k+m}{k}\right)^{\frac{\alpha}{2\beta'}} = C\left(1 + \frac{m-1+1}{k}\right)^{\frac{\alpha}{2\beta'}}
$$

$$
\leq C\left(1 + \frac{F(y, r) + 1}{k}\right)^{\frac{\alpha}{2\beta'}}
$$

$$
\leq C_\varepsilon \exp\left[\varepsilon C \left(\frac{F(y, r)}{k}\right)^{\frac{1}{(\beta-1)}}\right].
$$

Here we have to note that from Remark 3.11 it follows that $\beta > 1$, furthermore, from the conditions it follows that $\alpha, \beta' > 0$. The manipulation of the exponents used the trivial estimate $1 + x^{\frac{a}{a}} \leq \left(1 + x^{\frac{1}{a}}\right)^a$, where $x, a > 0$. As a result, by repeated application of (TC) we obtain that

$$
\sqrt{\frac{V(y, f(y, k))}{V(z, f(z, k))}} \leq C_\varepsilon \exp\left[\varepsilon C \left(\frac{F(y, r)}{k}\right)^{\frac{1}{(\beta-1)}}\right].
$$

∎

Proof [of Theorem 8.5] First of all, let us recall that $E \in W_0$ as it is clarified in Proposition 7.7. If $d(x,y) \le 2$, then the statement follows from (p_0). We use (8.48) with $r = \frac{1}{2} d(x,y)$ and start to handle the first term in

$$P_x\left(T_{x,r} < n\right) \max_{\substack{n \le k \le 2n \\ v \in \partial B(x,r)}} p_k\left(v, y\right).$$

Let us recall that from (TC) it follows that

$$\mathbb{P}(T_{x,r} < n) \le C \exp\left[-c k_x(n,r)\right], \tag{8.49}$$

and let us use $r \le d(v, y) \le 3r$ and (6.7) to get

$$P_x\left(T_{x,r} < n\right) \le C \exp\left[-c \left(\frac{E(x,r)}{n}\right)^{\frac{1}{\beta - 1}}\right].$$

Let us treat the second term. First we observe that

$$p_{2k+1}\left(y, v\right) \le \sum_{z \sim y} P_{2k}\left(y, z\right) P\left(z, v\right) \frac{1}{\mu(v)} \tag{8.50}$$

$$= \sum_{z \sim y} P_{2k}\left(y, z\right) P\left(v, z\right) \frac{1}{\mu(z)}$$

$$\le \max_{z \sim y} p_{2k}\left(y, z\right) \sum_{z \sim v} P\left(v, z\right)$$

$$= \max_{z \sim y} p_{2k}\left(y, z\right).$$

Let us recall that

$$p_{2k}\left(x, y\right) \le \sqrt{p_{2k}\left(x, x\right) p_{2k}\left(y, y\right)},$$

and by using the doubling properties of V and E it follows that for $w \sim v$ $d(y, v) \simeq d(y, w)$ (provided $v, w \ne y$)

$$\max_{\substack{n \le k \le 2n \\ w \in \partial B(x,r)}} p_k\left(w, y\right) \tag{8.51}$$

$$\le \max_{\substack{n \le 2k \le 2n \\ v \sim w \in \partial B(x,r)}} p_{2k}\left(v, y\right) \tag{8.52}$$

$$\le \max_{\substack{n \le 2k \le 2n \\ v \sim w \in \partial B(x,r)}} \frac{C}{\sqrt{V\left(y, e\left(y, 2k\right)\right) V\left(v, e\left(v, 2k\right)\right)}} \tag{8.53}$$

$$\le \max_{\substack{n \le 2k \le 2n \\ v \sim w \in \partial B(x,r)}} \frac{C}{V\left(y, e\left(y, n\right)\right)}.$$

Let us observe that $d(v, y) \le 3r + 2 \le 5r$ if $r \ge 1$, and let us apply Lemma 8.10 to proceed:

$$\max_{\substack{n \leq 2k \leq 2n \\ v \sim w \in \partial B(x,r)}} p_{2k}(v,y) P_x (T_{x,r} < n)$$

$$\leq \frac{C}{V(y, e(y,n))} C_\varepsilon \exp\left[\varepsilon C \left(\frac{E(y, 5r)}{n}\right)^{\frac{1}{(\beta-1)}} - c\left(\frac{E(x,r)}{n}\right)^{\frac{1}{(\beta-1)}}\right].$$

By choosing ε enough small and applying (TC), we have the inequality

$$\max_{\substack{n \leq k \leq 2n \\ v \in \partial B(x,r)}} p_k(v,y) \exp\left[-c\left(\frac{E(x, d(x,y))}{n}\right)^{\frac{1}{(\beta-1)}}\right]$$

$$\leq \frac{C}{V(y, e(y,n))} \exp\left[-c\left(\frac{E(x, d(x,y))}{n}\right)^{\frac{1}{(\beta-1)}}\right].$$

By symmetry, it follows that

$$p_{2n}(x,y) \leq C\left(\frac{1}{V(x, e(x,n))} + \frac{1}{V(y, e(y,n))}\right) \exp\left[-c\left(\frac{E(x, d(x,y))}{n}\right)^{\frac{1}{(\beta-1)}}\right]$$

$$= \frac{C}{V(x, e(x,n))}\left(1 + \frac{V(x, e(x,n))}{V(y, e(y,n))}\right) \exp\left[-c\left(\frac{E(x, d(x,y))}{n}\right)^{\frac{1}{(\beta-1)}}\right].$$

Now we use Lemma 8.10 to obtain

$$\frac{V(x, e(x,n))}{V(y, e(y,n))} \leq C_\varepsilon \exp\left[\varepsilon C \left(\frac{E(x, 2r)}{n}\right)^{\frac{1}{(\beta-1)}}\right],$$

and ε can be chosen to satisfy $\varepsilon C < \frac{c}{2}$ to receive

$$\left(1 + \exp\left[\left(\varepsilon C - \frac{c}{2}\right)\left(\frac{E(x, d(x,y))}{n}\right)^{\frac{1}{(\beta-1)}}\right]\right) \leq 2$$

$$p_{2n}(x,y) \leq \frac{2C}{V(x, e(x,n))} \exp\left[-\frac{c}{2}\left(\frac{E(x, d(x,y))}{n}\right)^{\frac{1}{(\beta-1)}}\right],$$

which is the needed estimate for even n. For the steps of an odd number the result follows by using, for $x \neq y$, the trivial inequality (8.50) and $d(x,y) \simeq d(x,z)$ if $z \neq x, y \sim z$. In particular, if the maximum in (8.50) is attained at $x = z$, then the statement follows from DUE and (p_0). ∎

Remark 8.2. With a slight modification of the beginning of the proof we can get

$$p_n(x,y) \leq \frac{C \exp[-ck_y(n,r)]}{V(x, e(x,n))} + \frac{C \exp[-ck_x(n,r)]}{V(y, e(y,n))}, \qquad (8.54)$$

where $r = \frac{1}{2}d(x,y)$ which is sharper than the above upper estimate. Let us note that our deduction shows that (8.54) is equivalent to the upper estimate.

8.8 Completion of the proof of Theorem 8.2

In this section we prove that each condition of Theorem 8.2 follows from the previous one. We will show that

$$PMV(E) \Longrightarrow wPMV(E) \Longrightarrow (MV) \Longrightarrow (MVG) \Longrightarrow (g_{0,1}).$$

The first two implications are trivial, so it is enough to deal with the rest.

Lemma 8.11. *If* $(p_0), (VD)$ *and* (TC) *hold, then*

$$(MV) \Longrightarrow (MVG) \Longrightarrow (g_{0,1}).$$

Proof   The implication $(MV) \Longrightarrow (MVG)$ follows from simply applying (MV) to $u(y) = g^U(y, w)$. On the other hand,

$$(MVG) \Longrightarrow (g_{0,1})$$

is immediate. If $d = d(x, y) > R$, then $g^{B(x,R)}(x, y) = 0$ and there is nothing to prove. Otherwise the extension of the summation, (VD) and (TC) imply the statement for $z \in B(x, d/2)$:

$$g_{0,1}^B(y, z) \le \frac{C}{V(x, d)} \sum_{v \in B(x,d)} g^B(y, v)\mu(v) \le \frac{C}{V(x, d)} E_y(x, R) \le C \frac{E(x, R)}{V(x, d)}.$$

∎

Theorem 8.6. *Assume,* (p_0) *and* (VD) *for* (Γ, μ). *Then for* $F \in W_0$

$$DUE(F) \Longrightarrow PMV(F).$$

Proof   Let us consider a Dirichlet solution $u_i(w) \ge 0$ on $B(x, R)$ with initial data $u_0 \in c_0(B(x, R))$. Let $F = F(x, R) = E(x, R)$. Consider $v_i(z) = \mu(z) u_i(z)$, $0 < c_1 < c_2 < c_3 < c_4$ and $n \in [c_3 F, c_4 F]$, $j \in [c_1 F, c_2 F]$. By definition,

$$u_n(w) \le \sum_{y \in \Gamma} P_{n-j}^B(w, y) u_j(y)$$

and

$$v_n(w) = \sum_{y \in B(x,R)} P_{n-j}^B(y, w) v_j(y) \le \mu(w) \max_{y \in B} p_{n-j}^B(y, w) \sum_{y \in B(x,R)} v_j(y),$$

from which by $p^B \le p$ and DUE, in particular by PUE we have, that

$$u_n(w) \le \max_{y \in B} p_{n-j}^B(w, y) \sum_{y \in B(x,R)} u_j(y)\mu(y) \le \frac{C}{V(w, f(w, n-j))} \sum_{y \in B(x,R)} u_j(y)\mu(y).$$

From using the doubling properties of f and V, it follows that $V(w, e(w, n - j)) \simeq V(x, R)$ and

$$u_n\left(w\right) \le \frac{C}{V\left(x,R\right)} \sum_{y\in B(x,R)} u_j\left(y\right)\mu\left(y\right). \qquad (8.55)$$

Finally, by summing (8.55) for $j \in [c_1F, c_2F]$, we obtain

$$u_n\left(w\right) \le \frac{C}{F\left(x,R\right)V\left(x,R\right)} \sum_{j=c_1F}^{c_2F} \sum_{y\in B(x,R)} u_j\left(y\right)\mu\left(y\right).$$

This means that this inequality holds for all $(n, w) \in [c_3F, c_4F] \times B\left(x, R\right) = \Psi$, and using the properties of V and F again, it holds for all $y \in V\left(x, R\right)$ and

$$\max_{\Psi} u \le \frac{C}{F\left(y,2R\right)V\left(y,2R\right)} \sum_{j=c_1F}^{c_2F} \sum_{y\in B(x,R)} u_j\left(y\right)\mu\left(y\right) \qquad (8.56)$$

is also satisfied. ∎

Proof   [of Theorem 8.2] Let us assume that the conditions (p_0), (VD), (TC) hold. From Theorem 8.3 and 8.4 it follows that

$$(g_{0,1}) \Longrightarrow (g_{\lambda,m}) \Longrightarrow DUE\left(E\right),$$

and $E \in W_0$. From Lemma 8.11 we know that $(MV) \Longrightarrow (g_{0,1})$. It is also clear that

$$DUE\left(E\right) \Longleftrightarrow PUE\left(E\right)$$

and $PMV\left(E\right)$ implies MV. Finally, we know that $E \in W_0$, and Theorem 8.6 provides $DUE \Longrightarrow PMV$. ∎

8.9 Upper estimates and the relative Faber-Krahn inequality

In this section another set of conditions, isoperimetric inequalities are given and we show their equivalence to diagonal upper estimates. Predecessors of these relative isoperimetric inequalities are discussed in Section 4.

Definition 8.8. *Let us write*

$$k\left(n, A, B\right) = \min_{z \in A} k_z\left(n, d\right), \tag{8.57}$$

where $d = d\left(A, B\right)$, *and*

$$\kappa\left(n, A, B\right) = \max\left\{k\left(n, A, B\right), k\left(n, B, A\right)\right\}. \tag{8.58}$$

Theorem 8.7. *Assume that* (Γ, μ) *satisfies* (p_0), *then the following inequalities are equivalent (assuming that each statement separately holds for all* $x, y \in \Gamma$, $R > 0, n > 0$, $D \subset A \subset B\left(x, 3R\right)$, $B = B\left(x, 2R\right)$ *with independent* $\delta, C > 0, \beta > 1$)

$$\overline{E}(A) \leq CE\left(x, R\right)\left(\frac{\mu\left(A\right)}{\mu\left(B\right)}\right)^{\delta}; \tag{FKE}$$

$$\lambda(A)^{-1} \leq CE\left(x, R\right)\left(\frac{\mu\left(A\right)}{\mu\left(B\right)}\right)^{\delta}; \tag{FK}$$

$$\rho(D, A)\mu\left(D\right) \leq E\left(x, R\right)\left(\frac{\mu\left(A\right)}{\mu\left(B\right)}\right)^{\delta}; \tag{FKρ}$$

$$p_n(x, x) \leq \frac{C}{V\left(x, e\left(x, n\right)\right)} \tag{8.59}$$

and $(VD) + (TC)$;

$$p_n\left(x, y\right) \leq \frac{C}{V\left(x, e\left(x, n\right)\right)} \exp\left[-c\frac{E\left(x, d\left(x, y\right)\right)^{\frac{1}{\beta-1}}}{n}\right] \tag{UE$\left(E\right)$}$$

and $(VD) + (TC)$,
where $e(x, n)$ *is the inverse of* $E\left(x, R\right)$ *in the second variable.*

Corollary 8.4. *If* (Γ, μ) *satisfies* $(p_0), (VD)$ *and* (TC), *then the following statements are equivalent (suppressing the quantors and assuming that the statements hold for all* $x \in \Gamma, R > 0, n > 0, D \subset A \subset B = B\left(x, 2R\right)$ *with independent* $\delta, \beta > 1$):

$$\frac{\overline{E}\left(A\right)}{\overline{E}\left(B\right)} \leq C\left(\frac{\mu\left(A\right)}{\mu\left(B\right)}\right)^{\delta}; \tag{8.60}$$

$$\frac{\lambda^{-1}(A)}{\lambda^{-1}(B)} \leq C \left(\frac{\mu(A)}{\mu(B)} \right)^{\delta}; \tag{8.61}$$

$$\frac{\rho(D, A)}{\rho(x, 2R)} \leq \left(\frac{\mu(A)}{\mu(D)} \right)^{\delta} \left(\frac{\mu(D)}{\mu(B)} \right)^{\delta-1}; \tag{8.62}$$

$$p_n(x, x) \leq \frac{C}{V(x, e(x, n))}; \tag{8.63}$$

$$p_n(x, y) \leq \frac{C}{V(x, e(x, n))} \exp\left[-c \left(\frac{E(x, d(x, y))}{n} \right)^{\frac{1}{\beta-1}} \right]. \tag{$UE(E)$}$$

Remark 8.3. By comparing the main result of the previous chapter to that of the present one, it can be immediately recognized that the mean value inequality is equivalent to the relative isoperimetric inequalities $(FKE), (FK), (FK\rho)$ and $(8.60 - 8.62)$ provided that (VD) and (TC) hold. In [50], a direct proof of $(MV) \Longrightarrow (FK)$ is given for measure metric spaces which works for weighted graphs as well.

Remark 8.4. Let us sketch the main implications. Under $(p_0), (VD)$ and (TC).

$$
\begin{array}{ccc}
(FK) & \leftrightarrow & DUE \\
\nearrow \quad \downarrow & & \uparrow \quad \searrow \\
(FK\rho) \longleftarrow (FKE) & UE & \longleftarrow PMV(E)
\end{array}
$$

8.9.1 Isoperimetric inequalities

In this section we prove and apply the following two statements to show that all the isoperimetric inequalities in Theorem 8.7 and Corollary 8.4 are equivalent. Let us recall that under the conditions $(p_0), (VD)$ and (TC),

$$\lambda^{-1}(x, 2R) \asymp E(x, 2R) \asymp \overline{E}(x, 2R) \asymp \rho(x, R, 2R) v(x, R, 2R) \tag{8.64}$$

holds.

Proposition 8.1. *The following statements are equivalent.*

$$\overline{E}(A) \leq C\mu(A)^{\delta} \text{ for all finite } A \subset \Gamma, \tag{8.65}$$

$$\lambda^{-1}(A) \leq C\mu(A)^{\delta} \text{ for all finite } A \subset \Gamma, \tag{8.66}$$

$$\rho(D, A)\mu(D) \leq C\mu(A)^{\delta} \text{ for all finite } D \subset A \subset \Gamma. \tag{8.67}$$

Proof [of Proposition 8.1] The implication $(8.65) \Longrightarrow (8.66)$ follows from Lemma 2.3, $(8.66) \Longrightarrow (8.67)$ from Theorem 3.1, and finally $(8.67) \Longrightarrow (8.65)$ from Proposition 4.1. ∎

Proposition 8.2. *The statements* $(FKE),(FK)$ *and* $(FK\rho)$ *are equivalent and so are* $(8.60),(8.61)$ *and* (8.62).

Proof   The statement follows from Proposition 8.1.   ■

Proposition 8.3. *Each of the statements* $(FKE),(FK)$ *and* $(FK\rho)$ *implies* (VD) *and* (TC).

Proof   First, let us observe that if one of them does so, then all of them do so, since they are equivalent by Proposition 8.2. So we can choose (FKE). Let $A = B(x, R)$, then $A = B(y, 2R)$ and we have immediately (VD) and (TC).   ■

Proposition 8.3 means that the volume and time comparison principles (VD) and (TC) can be set as preconditions in Theorem 8.7, as it is done in Corollary 8.4.

Proposition 8.4. *Theorem 8.7 and Corollary 8.4 mutually imply each other.*

Proof   According to Proposition 8.3, we can set (VD) and (TC) as preconditions, then using (7.3) the right hand side of each inequality $E(x, R)$ can be replaced by the needed term, thereby we receive that $(FKE) \Longrightarrow (8.60)$, $(FK) \Longrightarrow (8.61)$ and $(\rho) \Longrightarrow (8.62)$. The opposite implications can be seen by choosing $R' = \frac{3}{2}R$ and applying $(VD),(TC)$ and (8.1). This clearly gives the statement, if any of the isoperimetric inequalities is equivalent to the diagonal upper estimate, then all of them are.   ■

8.9.2 On-diagonal upper bound

In this section we shall show the following theorem which implies Theorem 8.7 according to Propositions 8.4 and (8.2).

Theorem 8.8. *If* (Γ, μ) *satisfies* $(p_0),(VD)$ *and* (TC), *then the following statements are equivalent:*

$$\lambda^{-1}(A) \le CE(x, R) \left(\frac{\mu(A)}{\mu(B)}\right)^{\delta} \quad \text{for all } A \subset B(x, 2R), \tag{8.68}$$

$$p_n(x, x) \le \frac{C}{V(x, e(x, n))}. \tag{8.69}$$

8.9.3 Estimate of the Dirichlet heat kernel

Remark 8.5. Let us recall Lemma 5.1. Given $C, a > 0$, and assuming that for any non-empty finite set $A \subset \Gamma$,

$$\lambda(A)^{-1} \le aC \left(\frac{1}{\mu(A)}\right)^{\delta}. \tag{8.70}$$

Let $f(x)$ be a non-negative function on Γ with finite support, then

$$a \|f\|_2^2 \left(\frac{\|f\|_2}{\|f\|_1} \right)^{2\delta} \leq C\mathcal{E}(f, f).$$

Now we have to make a careful detour as it was done in [27] and [48]. The strategy is the following. We consider the weighted graph Γ^* with the same vertex set as Γ with new edges and weights induced by the two-step transition operator $Q = P^2$

$$\mu_{x,y}^* = \mu(x) P_2(x, y).$$

If Γ^* is decomposed into two disconnected components, due to the periodicity of P the applied argument will work irrespective of which component is considered. We show that $(p_0), (VD), (TC)$ and (FK) hold on Γ^* if they hold on Γ. We deduce the Dirichlet heat kernel estimate for Q on Γ^*, then we show that it implies the same on Γ. We have to make this detour to ensure that

$$q(x, x) \geq c_0 > 0$$

holds for all $x \in \Gamma^*$ which will be needed in the proof of Lemma 8.14.

Lemma 8.12. *If $(p_0),(VD),(TC)$ and (FK) hold on Γ, then the same is true on Γ^*.*

Proof   The statement is evident for (p_0) and (VD). It is worth mentioning here that $\mu^*(x) = \mu(x)$, and from (2.11) we know that $\mu(x) \simeq \mu(y)$ if $x \sim y$. Let us observe that

$$B(x, 2R) \subset \overline{B}^*(x, R), \tag{8.71}$$
$$B^*(x, R) \subset B(x, 2R) \tag{8.72}$$

and

$$V^*(y, 2R) \leq V(y, 4R) \leq C^2 V(x, R) \tag{8.73}$$
$$\leq C^2 \mu\left(\overline{B}^*(x, R/2)\right) \leq C^2 V^*(x, R).$$

So the volumes of the above balls are comparable.

The next step is to show (TC).

$$E^* (y, 2R) = \sum_{z \in B^*(y,2R)} \sum_{k=0}^{\infty} Q_k^{B^*(y,2R)} (y, z)$$

$$\leq \sum_{z \in B(y,4R)} \sum_{k=0}^{\infty} P_{2k}^{B(y,4R)} (y, z)$$

$$\leq \sum_{z \in B(y,4R)} \sum_{k=0}^{\infty} P_{2k}^{B(y,4R)} (y, z) + P_{2k+1}^{B(y,4R)} (y, z)$$

$$= E (y, 4R) \leq C E (x, R/2)$$

$$= \sum_{z \in B(x,R/2)} \sum_{k=0}^{\infty} P_k^{B(x,R/2)} (x, z)$$

$$= \sum_{z \in B(x,R/2)} \sum_{k=0}^{\infty} P_{2k}^{B(x,R/2)} (x, z) + P_{2k+1}^{B(x,R/2)} (x, z) .$$

Now we use a trivial estimate.

$$P_{2k+1}^{B(x,R)} (x, z) = \sum_{w \sim z} P_{2k}^{B(x,R)} (x, w) P^{B(x,R)} (w, z) \qquad (8.74)$$

$$\leq \sum_{w \sim z} P_{2k}^{B(x,R)} (x, w) .$$

By summing up (8.74) for all z and recalling (2.12) which states $|\{w \sim z\}| \leq \frac{1}{p_0}$, we receive that

$$\sum_{z \in B(x,R/2)} P_{2k+1}^{B(x,R/2)} (x, z) \leq \sum_{z \in B(x,R/2)} \sum_{w \sim z} P_{2k}^{B(x,R/2)} (x, w)$$

$$\leq C \sum_{w \in \overline{B}(x,R/2)} P_{2k}^{B(x,R)} (x, w) .$$

As a result, we obtain that

$$E^* (y, 2R) \leq C \sum_{z \in \overline{B}(x,R/2)} \sum_{k=0}^{\infty} P_{2k}^{\overline{B}(x,R/2)} (x, z)$$

$$\leq C E^* (x, R/2 + 1) \leq C E^* (x, R) .$$

This shows that (TC) holds on Γ^*. We have also proved that

$$c E^* (x, R) \leq E (x, R) \leq C E^* (x, R) . \qquad (8.75)$$

It remains to show that from

$$\frac{1}{\lambda(A)} \leq C E (x, R) \left(\frac{\mu (A)}{V (x, R)} \right)^{\delta} \qquad (8.76)$$

it follows that

$$\frac{1}{\lambda^*(A)} \le CE^*(x, R) \left(\frac{\mu^*(A)}{V^*(x, R)}\right)^{\delta} \tag{8.77}$$

holds as well. Let us recall the inequality (5.17):

$$\lambda^*(A) \ge \lambda\left(\overline{A}\right) \tag{8.78}$$

By collecting the inequalities, we get the statement.

$$\frac{1}{\lambda^*(A)} \le \frac{1}{\lambda(\overline{A})} \le CE(x, R+1) \left(\frac{\mu\left(\overline{A}\right)}{V(x, R+1)}\right)^{\delta}$$
$$\le CE^*(x, R) \left(\frac{\mu^*(A)}{V^*(x, R)}\right)^{\delta}.$$

∎

In order to complete our proof we need the return from Γ^* to Γ. This is given in the following lemma.

Lemma 8.13. *Assume that* (Γ, μ) *satisfy* (p_0) (VD) *and* (TC). *In addition if* DUE *holds on* (Γ^*, μ), *then it holds on* $t(\Gamma, \mu)$.

Proof   From the statement

$$q_n(x, x) \le \frac{C}{V^*(x, e^*(x, n))}$$

and from the definition of q, (8.73) and (8.75) it follows that

$$p_{2n}(x, x) \le \frac{C}{V(x, e(x, 2n))}.$$

Finally for odd times the statement follows by a standard argument. From the spectral decomposition of $P_n^{B(x,R)}$ for finite balls we have that

$$P_{2n}^{B(x,R)}(x, x) \ge P_{2n+1}^{B(x,R)}(x, x)$$

and consequently,

$$p_{2n}(x, x) = \lim_{R \to \infty} p_{2n}^{B(x,R)}(x, x)$$
$$\ge \lim_{R \to \infty} p_{2n+1}^{B(x,R)}(x, x) = p_{2n+1}(x, x),$$

which gives the statement by using $(VD), (TC)$ and $e(x, n) \simeq e(x, n+1)$. ∎

Lemma 8.14. *If (p_0) is true and (FK) :*

$$\lambda(A)^{-1} \leq CE^*(x, R) \left(\frac{V^*(x, R)}{\mu(A)} \right)^{\delta} \tag{8.79}$$

holds for all $A \subset B^*(x, R)$ on (Γ^*, μ), then for all $x, y \in \Gamma$

$$q_n^{B^*(x,R)}(y, z) \leq \frac{C}{V^*(x, R)} \left(\frac{E^*(x, R)}{n} \right)^{1/\delta}.$$

Proof   The proof is a slight modification of the steps of proving $(a) \implies (b)$ in Theorem 5.3. We consider again the weighted graph Γ^* with the same vertex set as Γ with new weights induced by the two-step transition operator $Q = P^2$. Let $B = B^*(x, r)$ (omitting $*$ to simplify the notation)

$$f_n(z) = Q_n^B(z, y),$$
$$b_n = (f_n, f_n) = \|f_n\|_2^2 = q_{2n}^B(y, y).$$

We can see that

$$b_n - b_{n+1} = \frac{1}{2} \sum_{y, z \in B} (f(y) - f(z))^2 \mu_{y,z}^*.$$

From Lemma 8.12 we know that (FK) holds on Γ^* consequently from Lemma 5.1 (see Remark 8.5) and from $\|u_n\|_1 \leq 1$ it follows that

$$b_n - b_{n+1} \geq ca\|f\|_2^2 \left(\frac{\|f\|_2}{\|f\|_1} \right)^{2\delta} = cab_n^{1+\delta}, \tag{8.80}$$

where

$$a = \frac{V(x, R)^{\delta}}{E(x, R)}.$$

Now we use the inequality

$$\nu(x - y) \geq \frac{x^{\nu} - y^{\nu}}{x^{\nu-1} + y^{\nu-1}}, \tag{8.81}$$

which is true for all $x > y > 0$ and $\nu > 0$. Let $\nu = \frac{1}{\delta}$, $x = b_{n+1}^{-1/\nu}$ and $y = b_n^{-1/\nu}$ and apply (8.81) in (8.80):

$$\nu(b_{n+1}^{-1/\nu} - b_n^{-1/\nu}) \geq \frac{b_{n+1}^{-1} - b_n^{-1}}{b_{n+1}^{-(\nu-1)/\nu} + b_n^{-(\nu-1)/\nu}} = \frac{b_n - b_{n+1}}{b_{n+1}^{1/\nu} b_n + b_n^{1/\nu} b_{n+1}} \geq \frac{cab_n^{1+1/\nu}}{2b_n^{1+1/\nu}} = \frac{ca}{2},$$

whence

$$b_{n+1}^{-\delta} - b_n^{-\delta} \geq ac$$

and by summing up it for n, it follows that

$$b_n^{-\delta} \geq acn,$$

$$b_n \leq (acn)^{-\frac{1}{\delta}} .$$

This finally results in

$$q_{2n}^B (y, y) = b_n \leq \frac{C}{V (x, R)} \left(\frac{E (x, R)}{n} \right)^{1/\delta} ,$$

and the statement follows by using (8.45) for q_n^B ■

Now we consider the following path decompositions.

Lemma 8.15. *Let $p_n (x, y)$ be the heat kernel of a random walk on an arbitrary weighted graph (Γ, μ). Let $A \subset \Gamma, x, y \in \Gamma, n > 0$, then*

$$p_n (x, y) \leq p_n^A (x, y) + P_x (T_A < n) \max_{\substack{z \in \partial A \\ 0 \leq k < n}} p_k (z, y) , \qquad (8.82)$$

$$p_n (x, y) \leq p_n^A (x, y) + P_x (T_A < n/2) \max_{\substack{z \in \partial A \\ n/2 \leq k < n}} p_k (z, y) \qquad (8.83)$$

$$+ P_y (T_A < n/2) \max_{\substack{z \in \partial A \\ n/2 \leq k < n}} p_k (z, x) . \qquad (8.84)$$

Proof   Both inequalities follow from the first exit decomposition starting from x or from x and y as well. The proof is left to the reader as an exercise. ■

8.9.4 Proof of the diagonal upper estimate

Proof   [of Theorem 8.8] First we show the implication $(FK) \Longrightarrow (DUE)$ on Γ^* assuming $(p_0), (VD)$ and (TC). We follow the main lines of [40]. Let us choose r so that $Ln = E (x, r)$ for a large $L > 0$. From (8.83) we have that for $B = B^* (x, r)$

$$q_n (x, x) \leq q_n^B (x, x) + 2Q_x (T_B < n/A) \max_{\substack{z \in \partial B \\ n/A \leq k < n}} q_k (z, x) . \qquad (8.85)$$

From (8.45) it follows that for all $n/A \leq k < n$

$$q_k (z, x) \leq \sqrt{q_k (z, z) q_k (x, x)} \leq \max_{v \in \overline{B}} q_k (v, v) \leq C_1 \max_{v \in \overline{B}} q_{\lfloor n/A \rfloor} (v, v) .$$

By substituting this into (8.85) we obtain for $x_1 \in \overline{B}$ that

$$q_n (x, x) \leq q_n^B (x, x) + 2Q_x (T_B < n/A) C_1 q_{\lfloor n/A \rfloor} (x_1, x_1) . \qquad (8.86)$$

We continue this procedure. In the ith step we have

$$q_{n_{i-1}}(x_i, x_i) \le q_{n_i}^{B_i}(x_i, x_i) + 2Q_{x_i}(T_{B_i} < n_{i+1}) C_1 q_{n_i}(x_{i+1}, x_{i+1}), \quad (8.87)$$

where $n_i = \lfloor n/A^i \rfloor, r_i = e(x_i, Ln_i), B_i = B(x_i, r_i), x_{i+1} \in \overline{B}_i$. Let $m = \lfloor \log_A n \rfloor$ and let us stop the iteration at m. This means that $1 \le \lfloor \frac{n}{A^m} \rfloor = n_m < A$. By the definition of n_i, r_i and from (TC) it follows that

$$A = \frac{Ln_i}{Ln_{i+1}} = \frac{E(x_i, r_i)}{E(x_{i+1}, r_{i+1})} \le \frac{E(x_i, 2r_i)}{E(x_{i+1}, r_{i+1})} \le C\left(\frac{2r_i}{r_{i+1}}\right)^\beta,$$

which results in

$$r_{i+1} \le \sigma r_i \qquad (8.88)$$

if $\sigma = 2\left(\frac{C}{A}\right)^{\frac{1}{\beta}} < 1$ and $A > 2^\beta C$. From Lemma 8.14 the first term can be estimated as follows

$$q_{n_i}^{B_i}(x_i, x_i) \le \frac{C}{V(x_i, r_i)}\left(\frac{E(x, r_i)}{n_i}\right)^{1/\delta}$$

$$= \frac{C}{V(x_i, r_i)} L^{1/\delta} = \frac{CL^{1/\delta}}{V(x, 2r)} \frac{V(x, 2r)}{V(x_i, r_i)}$$

$$\frac{CL^{1/\delta}}{V(x, R)}\left(\frac{2}{\sigma}\right)^{\alpha i}.$$

Let us observe that by definition of $k = k_{x_i}(n_{i+1}, r_i)$

$$\frac{n_{i+1}}{k+1} > q \min_{y \in B_i} E\left(y, \frac{r_i}{k+1}\right),$$

but from (TC) we obtain

$$Ln_i = E(x_i, r_i) \le CE(y, r_i) \le C(k+1)^\beta E\left(y, \frac{r_i}{k+1}\right)$$

$$\le \frac{C}{q}(k+1)^{\beta-1} n_i,$$

which results in

$$Q_{x_i}(T_B < n_{i+1}) \le C \exp\left[-ck_{x_i}(n_{i+1}, r_i)\right] \le C \exp\left[-c\left(\frac{\sigma}{C}L\right)^{\frac{1}{\beta-1}}\right].$$

This means that

$$Q_{x_i}(T_B < n_{i+1}) \le \frac{\varepsilon}{2}$$

if a large enough L is chosen. Inserting this in (8.87), we get

$$q_{n_{i-1}}(x_i, x_i) \le \frac{CL^{1/\delta}}{V(x, R)}\left(\frac{2}{\sigma}\right)^{\alpha i} + \varepsilon C_1 q_{n_i}(x_{i+1}, x_{i+1}). \qquad (8.89)$$

By summing up (8.89) the iteration results in

$$q_n\left(x,x\right) \leq \frac{CL^{1/\delta}}{V\left(x,R\right)} \sum_{i=1}^{m} \left(\left(\frac{2}{q}\right)^{\alpha} \varepsilon\right)^i + \left(\varepsilon C_1\right)^m q_m\left(x_m,x_m\right). \qquad (8.90)$$

By choosing L large enough, $\varepsilon < \min\left\{\left(\frac{q}{2}\right)^{\alpha}, \frac{1}{C_1}\right\}$ can be ensured. This yields that the sum in the first term is bounded by $1/\left(1 - \varepsilon\left(\frac{2}{\sigma}\right)^{\alpha}\right) < C$. The second term can be treated as follows.

$$q_m\left(x_m,x_m\right) = \frac{1}{\mu\left(x_m\right)} Q_m\left(x_m,x_m\right) \leq \frac{1}{\mu\left(x_m\right)}.$$

From (8.88) we have that

$$d\left(x,x_m\right) \leq r + r_1 + ... + r_m \leq r \sum_{i=0}^{m} \sigma^i < r \frac{1}{1-\sigma} = Cr,$$

$$\frac{1}{\mu\left(x_m\right)} = \frac{1}{V\left(x,r\right)} \frac{V\left(x,r\right)}{V\left(x,Cr\right)} \frac{V\left(x,Cr\right)}{\mu\left(x_m\right)}$$

$$\leq \frac{1}{V\left(x,R\right)} \left(Cr\right)^{\alpha}.$$

Consequently, we are ready if

$$\left(Cr\right)^{\alpha} \varepsilon^m < C'.$$

Let us remark that $E\left(z,R\right) \geq R$. This implies that $e\left(x,n\right) \leq n$. From the definition of m and $E\left(x,r\right) = Ln$,

$$\left(Cr\right)^{\alpha} \varepsilon^m \leq \left(Cr\right)^{\alpha} \varepsilon^{\log_A n} = \left[CE\left(x,r\right)\right]^{\alpha} n^{\log_A \varepsilon} = \left(CL\right)^{\alpha} n^{\alpha+\log_A \varepsilon} \leq C$$

if $\varepsilon < A^{-\alpha}$, and L is large enough. Finally from (8.90) we receive that

$$q_n\left(x,x\right) \leq \frac{CL^{1/\delta}}{V\left(x,R\right)} \sum_{i=1}^{m} \left(\left(\frac{2}{q}\right)^{\alpha} \varepsilon\right)^i + \left(\varepsilon C_1\right)^m q_{n_m}\left(x_m,x_m\right) \qquad (8.91)$$

$$\leq \frac{C}{V\left(x,R\right)} = \frac{C}{V\left(x,e\left(x,Ln\right)\right)} \leq \frac{C}{V\left(x,e\left(x,n\right)\right)}, \qquad (8.92)$$

if $\varepsilon < \min\left\{\left(\frac{q}{2}\right)^{\alpha}, \frac{1}{C_1}, A^{-\alpha}\right\}$ absorbing all the constants in C. Lemma (8.13) shows that DUE holds on Γ as well. ∎

8.9.5 Proof of $DUE\left(E\right) \Longrightarrow \left(FK\right)$

In this section we derive the relative or local form of the Faber-Krahn inequality. In [27] the equivalence of the relative Faber-Krahn inequality:

$$\lambda(A) \geq \frac{c}{R^2} \left[\frac{V(x,R)}{\mu(A)} \right]^{2/\alpha} \quad \text{for all } A \subset \Gamma \tag{8.93}$$

and the conjunction of (VD) and (UE_2):

$$p_n(x,y) \leq \frac{C}{V(x,\sqrt{n})} \exp\left[-C\frac{d(x,y)^2}{n} \right] \tag{8.94}$$

was shown. This is possible thanks to the fact (cf. Lemma 2.2) that the inequality of opposite orientation

$$\lambda(x,R) \leq CR^{-2}$$

is true if (VD) holds. Of course, this is too weak in the region $\beta > 2$, where, for instance, in the case of strongly recurrent graphs with $V(x,R) \simeq R^\alpha, E(x,R) \simeq R^\beta$ and (H), the smallest eigenvalue satisfies

$$\lambda(x,R) \simeq R^{-\beta}.$$

Proposition 8.5. *For all (Γ, μ), the conditions (VD) and $DUE(E)$ imply the relative Faber-Krahn inequality (FK):*

$$\lambda^{-1}(A) \leq CE(x,R) \left(\frac{\mu(A)}{\mu(B)} \right)^\delta \quad \text{for all } A \subset B(x,R). \tag{8.95}$$

Proof The proof works along the lines of [27]. We start with

$$P_k(x.x) \leq \frac{c\mu(x)}{V(x,e(x,k))}. \tag{8.96}$$

Let $\varnothing \neq A \subset \Gamma$ and $A \subset B(x,R)$. If $\lambda \geq 1/2$, the proof is immediate,

$$2E(x,R) \left(\frac{\mu(A)}{\mu(B)} \right)^\delta \geq 2E(x,R) \left(\frac{\mu(x)}{V(x,R)} \right)^\delta \geq 2R^{2-\alpha\delta} \geq 2,$$

if $\delta \leq \frac{2}{\alpha}$. Assume that $\lambda < 1/2$. Let us consider the trace of P_k^A

$$tr P_k^A = \sum_{y \in A} P_k^A(y,y) \leq \frac{C\mu(A)}{\inf\limits_{y \in A} V(y,e(y,k))},$$

$$\lambda_{\max}(P^A)^k \leq \lambda_{\max}(P_k^A) \leq tr P_k^A,$$

whence

$$\lambda_{\max}(P^A) \leq \left(\frac{C\mu(A)}{\inf\limits_{y \in A} V(y,e(y,k))} \right)^{1/k}.$$

Using $1 - \xi \geq \frac{1}{2} \log(\frac{1}{\xi})$ for $\xi \in [\frac{1}{2}, 1]$, it follows for $\xi = \lambda_{\max}(P^A)$ that

$$\lambda(A) = 1 - \lambda_{\max}(P^A) \geq \frac{1}{2k} \log \frac{\inf\limits_{x \in A} V(x, e(x, k))}{C\mu(A)}.$$

Let us start with

$$k_1 = E\left(x, \left\lceil R\left(\frac{\mu(A)}{V(x, R)}\right)^{\frac{1}{\alpha}}\right\rceil\right).$$

By this definition, for $z \in A \subset B(x, R)$,

$$k_1 \leq E(x, R) \leq CE(z, R) \tag{8.97}$$

and consequently

$$e(z, k_1) \leq CR,$$

while

$$e(x, k_1) \geq R\left(\frac{\mu(A)}{V(x, R)}\right)^{\frac{1}{\alpha}}$$

$$\geq cR\left(\frac{\mu(A)}{V(z, R)}\right)^{\frac{1}{\alpha}}.$$

We obtain that

$$\frac{\mu(A)}{V(z, e(z, k_1))} = \frac{V(z, R)}{V(z, e(z, k_1))} \frac{\mu(A)}{V(z, R)}$$

$$\leq C\left(\frac{R}{e(z, k_1)}\right)^{\alpha} \left(\frac{Ce(x, k_1)}{R}\right)^{\alpha} \leq C,$$

which shows that $\frac{\inf\limits_{y \in A} V(y, e(y, k))}{\mu(A)} \geq \varepsilon$. The increase of the constant $k = Ck_1$ makes this ratio larger, and then e and $R\left(\frac{\mu(A)}{V(z, R)}\right)^{\frac{1}{\alpha}}$ can be separated from zero.   ∎

8.9.6 Generalized Davies-Gaffney inequality

In this section we give a discrete proof of the generalization of the Davies-Gaffney inequality. In [27], a discrete proof was given with a single exception of a continuous argument for the diffusive case (Gafney's Lemma, Lemma 5.1 of [27]) .

Let us recall the notation

$$k(n, A, B) = \min_{z \in A} k_z(n, d), \tag{8.98}$$

where $d = d(A, B)$ and

$$\kappa(n, A, B) = \max\{k(n, A, B), k(n, B, A)\}. \tag{8.99}$$

To simplify the notation, let us write $\kappa(n, x, y) = \kappa(n, \{x\}, \{y\})$.

Theorem 8.9. *If $(\overline{E})$ holds for a reversible Markov chain, then there is a constant $c > 0$ such that for all $A, B \subset V$, the Davies-Gaffney type inequality* (DG)

$$\sum_{x \in A, y \in B} p_n(x, y)\mu(x)\mu(y) \leq [\mu(A)\mu(B)]^{1/2} \exp[-c\kappa(n, A, B)] \qquad (8.100)$$

holds.

Proof   Using the Cauchy-Schwarz inequality, we obtain

$$\sum_{x \in A, y \in B} P_n(x, y)\mu(x) \qquad (8.101)$$

$$= \sum_{x \in \Gamma} \mu(x)^{1/2} I_A(x) \left[\mu^{1/2}(x) I_A(x) \sum_{y \in B} P_n(x, y) I_B(y) \right] \qquad (8.102)$$

$$\leq (\mu(A))^{1/2} \left\{ \sum_{x \in \Gamma} \mu(x) I_A(x) \left[\sum_{y \in \Gamma} P_n(x, y) I_B(y) \right]^2 \right\}^{1/2}.$$

Let us deal with the second term writing $r = d(A, B)$

$$\sum_{x \in \Gamma} \mu(x) I_A(x) \left[\sum_{y \in \Gamma} P_n(x, y) I_B(y) \right]^2 \qquad (8.103)$$

$$= \sum_{x \in \Gamma} \mu(x) I_A(x) \sum_{y \in \Gamma} P_n(x, y) I_B(y) \sum_{z \in \Gamma} P_n(x, z) I_B(z)$$

$$= \sum_{x \in \Gamma} \sum_{y \in \Gamma} \sum_{z \in \Gamma} P_n(x, z) I_B(z) \mu(x) I_A(x) P_n(x, y) I_B(y) \qquad (8.104)$$

$$= \sum_{z \in \Gamma} \sum_{y \in \Gamma} \sum_{x \in \Gamma} P_n(z, x) I_B(z) \mu(z) I_A(x) P_n(x, y) I_B(y) \qquad (8.105)$$

$$\leq \sum_{z \in \Gamma} \sum_{x \in \Gamma} P_n(z, x) I_B(z) \mu(z) I_A(x) \sum_{y \in \Gamma} P_n(x, y) I_B(y)$$

$$\leq \sum_{z \in \Gamma} \sum_{x \in \Gamma} P_n(z, x) I_B(z) \mu(z) I_A(x) \qquad (8.106)$$

$$\leq \sum_{z \in \Gamma} P_n(z, A) I_B(z) \mu(z) \leq \sum_{z \in \Gamma} P_z(T_{z,r} < n) I_B(z) \mu(z)$$

$$\leq \max_{z \in B} \exp[-ck(z, n, r)] \mu(B). \qquad (8.107)$$

The combination of (8.101) and (8.103) gives the second term in the definition of κ and by symmetry we can obtain the first one. ∎

8.9.7 Off-diagonal upper estimate

In this section we present an upper estimate of the heat kernel.

Proposition 8.6. *Assume that (Γ, μ) satisfies (p_0), $PMV(E)$ and (TC). Let $x, y \in \Gamma$, then there are $c, C > 0, \beta > 1$ such that for all $x, y \in \Gamma, n > 0$*

$$p_n(x, y) \leq \frac{C}{\sqrt{V(x, e(x, n)) V(y, e(y, n))}} \exp\left[-c\left(\frac{E(x, d(x, y))}{n}\right)^{\frac{1}{\beta-1}}\right].$$

$$(8.108)$$

Proof   The proof combines the repeated use of the parabolic mean value inequality and the Davies-Gaffney inequality. We follow the idea of [70]. Write $R = e(x, n)$, $S = e(y, n)$ and assume that $d \geq \frac{2}{3}(R + S)$ which ensures that $r = d - R - S \geq \frac{1}{3}d$. From $PMV(E)$ it follows that

$$p_n(x, y) \leq \frac{C}{V(x, R) E(x, R)} \sum_{c_1 E(x,R)}^{c_2 E(x,R)} \sum_{z \in V(x,R)} p_k(z, y) \mu(z)$$

and using (PMV) for $p_k(z, y)$ we get

$$p_n(x, y) \leq \frac{C}{V(x, R) V(y, S) n^2} \sum_{i=c_1 n}^{c_2 n} \sum_{z \in V(x,R)} \sum_{j=c_1 n+i}^{c_2 n+i} \sum_{w \in V(y,S)} p_j(z, w) \mu(z) \mu(w)$$

$$(8.109)$$

Now by (DG) and (6.7) and writing $A = B(x, R), B = B(y, S)$, we obtain

$$p_n(x, y) \leq \frac{C\sqrt{V(x, R) V(y, S)}}{V(x, R) V(y, S) n^2} \sum_{i=c_1 n}^{c_2 n} \sum_{j=c_1 n+i}^{c_2 n+i} e^{-c\kappa(n, A, B)}.$$

$$(8.110)$$

Using (TC) and $R < \frac{3}{2}d$, we can see that

$$\max_{z \in V(x,R)} \exp\left[-c\left(\frac{E(z, d/3)}{n}\right)^{\frac{1}{\beta-1}}\right] \leq \exp\left[-c\left(\frac{E(x, d)}{n}\right)^{\frac{1}{\beta-1}}\right],$$

and similarly

$$\max_{w \in V(y,R)} \exp\left[-c\left(\frac{E(w, d/3)}{n}\right)^{\frac{1}{\beta-1}}\right] \leq \exp\left[-c\left(\frac{E(y, d)}{n}\right)^{\frac{1}{\beta-1}}\right]$$

$$\leq \exp\left[-c\left(\frac{E(x, d)}{n}\right)^{\frac{1}{\beta-1}}\right],$$

which results in

$$p_n(x, y) \leq \frac{C}{\sqrt{V(x, R) V(y, S)}} \exp\left[-c\left(\frac{E(x, d)}{n}\right)^{\frac{1}{\beta-1}}\right].$$

It remains to treat the case $d(x,y) < \frac{2}{3}(R+S)$. In this case $\kappa(n, B(x,R), B(y,S)) = 1$ in (8.110). On the other hand either $d(x,y) < R$ or $d(x,y) < S$

$$E(x,d) \le E(x, e(x,n)) = n,$$

which results in $1 \le C \exp\left[-c\left(\frac{E(x,d)}{n}\right)^{\frac{1}{\beta-1}}\right]$ for a fixed $C > 0$, and similarly,

if $d(x,y) < S$, $1 \le C \exp\left[-c\left(\frac{E(y,d)}{n}\right)^{\frac{1}{\beta-1}}\right] \le C \exp\left[-c\left(\frac{E(x,d)}{n}\right)^{\frac{1}{\beta-1}}\right]$ which

gives the statement. ∎

Theorem 8.10. *Assume that* (Γ, μ) *satisfies* (p_0), (VD), (TC) *and* $DUE(E)$, *then* $UE(E)$ *holds:*

$$p_n(x,y) \le \frac{C}{V(x, e(x,n))} \exp\left[-c\left(\frac{E(x, d(x,y))}{n}\right)^{\frac{1}{\beta-1}}\right]. \qquad (8.111)$$

Proof   From Theorem 8.6 we have that

$$DUE(E) \Longrightarrow PMV(E).$$

Now we can use Proposition 8.6 which states that from $PMV(E)$ and (TC) it follows that

$$p_n(x,y) \le \frac{C}{\sqrt{V(x, e(x,n)) V(y, e(y,n))}} \exp\left[-c\left(\frac{E(x, d(x,y))}{n}\right)^{\frac{1}{\beta-1}}\right].$$

Let us use Lemma 8.10:

$$p_n(x,y) \le \frac{C}{V(x, e(x,n))} \sqrt{\frac{V(x, e(x,n))}{V(y, e(y,n))}} \exp\left[-c\left(\frac{E(x, d(x,y))}{n}\right)^{\frac{1}{\beta-1}}\right]$$

$$\le \frac{CC_\varepsilon}{V(x, e(x,n))} \exp\left[\varepsilon C\left(\frac{E(x,r)}{n}\right)^{\frac{1}{(\beta-1)}} - c\left(\frac{E(x, d(x,y))}{n}\right)^{\frac{1}{\beta-1}}\right].$$

and choosing ε enough small, we get the statement. ∎

9

Lower estimates

In the previous chapters we have shown the upper estimate

$$p_n(x,y) \leq \frac{C}{V(x,e(x,n))} \exp\left[-c\left(\frac{E(x,d(x,y))}{n}\right)^{\frac{1}{\beta-1}}\right].$$

Now we are interested in similar lower estimates.

First we deal with two important intermediate estimates, the near diagonal lower estimate and the particular lower estimate. The latter one is also a near diagonal lower estimate but it is for Dirichlet heat kernels on balls. We will give a condition which is equivalent to this one. Then we will show the particular lower estimate if the elliptic Harnack inequality holds.

Definition 9.1. *The particular lower estimate $PLE(F)$ with respect to a function F holds if there are $\varepsilon, \delta, c > 0$ constants such that for all $x, y \in \Gamma, n \geq R > 0$*

$$\widetilde{p}_n^{B(x,R)}(x,y) \geq \frac{c}{V(x,f(x,n))} \tag{9.1}$$

provided that $d(x,y) \leq \delta f(x,n), n \leq \varepsilon F(x,R)$.

9.1 Parabolic super mean value inequality

We define the parabolic super mean value inequality $PSMV$, and we show that it implies a local lower bound.

Definition 9.2. *We say that (the strong form of) the parabolic super mean-value inequality $\mathbf{PSMV}(F)$ holds on (Γ, μ) with respect to a function F if there is a $0 < \varepsilon < 1$ such that for any fixed constants $0 < c_1 < c_2 < c_3 < c_4 \leq c_5$, $c_4 - c_1 < \varepsilon$, there are $\delta, c > 0$ such that for arbitrary $x \in \Gamma$ and $R > 0$, using the notations $F = F(x,R)$, $B = B(x,R), \mathcal{D} = [0, c_5 F] \times B$, $A^+ = [c_3 F, c_4 F] \times B(x, \delta R)$, $A^- = [c_1 F, c_2 F] \times B(x, \delta R)$ for any non-negative Dirichlet super-solution of the heat equation*

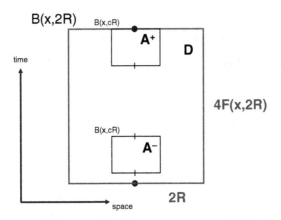

Fig. 9.1. The test balls for $PSMV$

$$\triangle^B u \leq \partial_n u_n$$

on $\mathcal{D}$, the inequality

$$\min_{A^+} \widetilde{u}_k \geq \frac{c}{\nu(A^-)} \sum_{(i,y)\in A^-} \widetilde{u}_i(y)\mu(y) \tag{9.2}$$

holds. (See Figure 9.1.)

Definition 9.3. *We say that under the conditions in Definition 9.2 the weak parabolic super mean value inequality* **wPSMV** (F) *holds if*

$$\widetilde{u}_n(x) \geq \frac{c}{\nu(A^-)} \sum_{(i,y)\in A^-} \widetilde{u}_i(y)\mu(y), \tag{9.3}$$

where $n = c_4 F(x, R)$.

Remark 9.1. It is evident that $PSMV$ implies the weaker form $wPSMV$.

Exercise 9.1. Show that $wPSMV$ implies $PSMV$ as well.

Following Sung's idea [92], we derive that the $PSMV(F)$ is equivalent to the particular lower estimate $PLE(F)$, which plays a crucial role in the proof of the lower estimate, as it will be shown below and in Chapter 13.

Theorem 9.1. *Assume that a weighted graph* (Γ, μ) *satisfies* (p_0) *and* (VD), *then the following statements are equivalent.*
1. there is an $F \in W_0$ *for which* $PSMV(F)$ *holds,*
2. there is an $F \in W_0$ *for which* $PLE(F)$ *holds.*

Lemma 9.1. *The estimate* $PLE\,(F)$ *implies a slightly stronger version of it: there are* $\delta, \varepsilon > 0$ *and* $c > 0, n \geq R > 0$ *such that*

$$p_m^{B(x,R)}(y, z) + p_{m+1}^{B(x,R)}(y, z) \geq \frac{c}{V(x, f(x, m))} \tag{9.4}$$

holds provided that $y, z \in B\,(x, r)$, *where* $r = \delta/2 f(x, m)$, *and* $m \leq \varepsilon F = \varepsilon F(x, R)$.

Exercise 9.2. The proof of the lemma is left to the reader as an exercise.

Proof   [of Theorem 9.1] For the super-solution $\widetilde{u}$, we have that for all $c_3 F \leq k \leq c_4 F, c_1 F \leq i \leq c_2 F$,

$$\widetilde{u}_k\,(y) \geq \sum_{z \in B(x,R)} \widetilde{p}_{k-i}^{B(x,R)}(y, z) \mu\,(z)\, u_i\,(z).$$

In order to use (9.4) we choose $\delta^* = \frac{1}{C_F}\,(c_3 - c_2)^{\frac{1}{\beta'}}\,\delta/2$ and $r = \delta^* R$. From the condition $c_4 - c_1 \leq \varepsilon$, it follows that $k - i \leq \varepsilon F\,(x, R)$ is satisfied and $PLE\,(F)$ can be applied:

$$\widetilde{u}_k\,(y) \geq \sum_{y \in B(x, \delta^* R)} \widetilde{p}_{k-i}^{B}(y, z) \mu\,(z)\, u_i\,(z)$$

$$\geq \frac{c}{V(x, f\,(x, k - i))} \sum_{y \in B(x, \delta^* R)} \mu\,(y)\, u_i\,(y).$$

Now let us sum this inequality for $i : c_1 F \leq i \leq c_2 F$ and divide it by $(c_2 - c_1)\,F$ to obtain

$$\widetilde{u}_k\,(y) \geq \frac{c}{F\,(x, R)} \sum_{i=c_1 F}^{c_2 F} \frac{1}{V(x, f\,(x, k))} \sum_{y \in B(x, \delta^* R)} \mu\,(y)\, u_i\,(z)$$

$$\geq \frac{c}{V\,(x, R)\, F\,(x, R)} \sum_{i=c_1 F}^{c_2 F} \sum_{y \in B(x, \delta^* R)} \mu\,(y)\, u_i\,(z).$$

Now we close the circle by proving $PSMV\,(F) \implies PLE\,(F)$. We apply $PSMV$ twice.

1. Let ε, δ, c_i be the constants in $PSMV\,(F)$, c_is will be specified later (c_is determine δ as well), furthermore $F = F\,(p, R), B = B\,(p, R)$, $r_1 = R/8, F_1 = F\,(p, r_1), D_1 = B\,(p, \delta r_1), m = c_2 F_1$ and $D = B\left(p, \delta \frac{R}{4}\right)$. Let us define

$$u_n\,(y) = \begin{cases} \sum_{z \in D} \widetilde{p}_{n-m}^{B}\,(y, z)\, \mu\,(z) & \text{if } n > m \\ \\ 1 & \text{if } n \leq m \end{cases}.$$

This is a solution on $D \times [0, \infty]$ of the equation

$$Pu_n = u_{n+1}$$

and $u_n \geq 0$. From $PSMV(F)$ it follows that

$$u_k(x) \geq \frac{c}{V(x, \delta r_1) F_1} \sum_{i-c_1 F_1}^{c_2 F_1} \sum_{w \in D_1} \mu(w) \tilde{u}_i(w)$$

provided that $x \in B_1$ and $c_3 F_1 < k < c_4 F_1$. From the definitions of u_k, (VD) and $F \in W_0$ it follows that

$$\tilde{u}_k(x) \geq \frac{c}{V(x, \delta r_1) F_1} \sum_{i=c_1 F_1}^{c_2 F_1} \sum_{w \in D_1} c\mu(w) \geq c.$$

Again from the definition of u_k and $D_1 \subset D$, $c_3 F_1 < k < c_4 F_1$, $x \in D_1$ it follows that

$$\sum_{z \in D} \tilde{p}^B_{k-m}(x, z) \mu(z) \geq c \tag{9.5}$$

or equivalently,

$$\sum_{z \in D} \tilde{p}^B_i(x, z) \mu(z) \geq c \tag{9.6}$$

if $x \in D_1, (c_3 - c_2) F_1 < i < (c_4 - c_2) F_1$.

2. We will use the parabolic super mean value inequality in a new ball B_2 for $\tilde{p}^B_i(x, y)$ with the same set of constants c_i, hence with the same δ as well. Let $r_2 = R/2$ in $B_2 = B(x, r_2)$, $D_2 = B(x, \delta r_2)$, $F_2 = F(x, r_2)$. We apply $PSMV$ in B_2 and obtain that for $c_3 F_2 < l < c_4 F_2$, $y \in D_2$

$$\tilde{p}^B_l(x, y) \geq \frac{c}{V(x, \delta r_2) F(x, r_2)} \sum_{i=c_1 F_2}^{c_2 F_2} \sum_{z \in D_2} \tilde{p}^B_i(x, z) \mu(z)$$

if, in addition, $B_2 \subset B(p, R)$. Let $x \in B\left(p, \frac{\delta R}{8}\right)$. This ensures that $B_2 \subset B(p, R)$ and $D_2 \supset D$, and for $y \in B\left(x, \frac{\delta R}{4}\right) \subset D_2$, $\left(B\left(x, \frac{\delta R}{4}\right) \subset B\left(p, \frac{\delta R}{2}\right)\right.$ as well) we obtain that

$$\tilde{p}^B_l(x, y) \geq \frac{c}{V(x, \delta r_2) F(x, r_2)} \sum_{cF_2}^{cF_2} \sum_{z \in D_2} \tilde{p_i}^B(y, z) \mu(z). \tag{9.7}$$

In order to use (9.6), we need that

$$c_2 F_2 \geq (c_4 - c_2) F_1, \tag{9.8}$$

and

$$c_1 F_2 \leq (c_3 - c_2) F_1. \tag{9.9}$$

From the assumption $F \in W_0$, it follows that (9.8) is satisfied if

$$c_4 = c_2 \left(1 + c_F 4^{\beta'} \right),$$

and (9.9) is satisfied if, for a $0 < q < 1$,

$$c_1 = q \frac{(c_3 - c_2)}{C_F} 4^{-\beta}$$

for any $0 < q < 1$. Finally,

$$0 < c_4 - c_1 = c_2 \left(1 + c_F 4^{\beta'} \right) - \frac{(c_3 - c_2)}{C_F} 4^{-\beta} < \varepsilon$$

and $c_1 < c_2 < c_3 < c_4$ can be ensured with an appropriate choice of c_2, c_3 and q. Using (9.8) and (9.9) and $D_2 \supset D$, the estimate in (9.7) can be continued as follows:

$$\widetilde{p}_l^B (x, y)$$

$$\geq \frac{c}{V (x, \delta r_2) F (x, r_2)} \sum_{c_1 F_2}^{c_2 F_2} \sum_{z \in D_2} \widetilde{p}_i^{\ B} (y, z) \mu (z)$$

$$\geq \frac{c}{V (x, \delta r_2) F (x, r_2)} \sum_{(c_3 - c_2) F_1}^{(c_4 - c_2) F_1} \sum_{z \in D} \widetilde{p}_i^B (y, z) \mu (z).$$

Now we apply (9.6) to deduce

$$\widetilde{p}_l^{B(p,R)} (x, y) \geq \frac{c}{V (x, \delta r_2) F (x, r_2)} \sum_{(c_3 - c_2) F_1}^{(c_4 - c_2) F_1} c$$

$$\geq \frac{c}{V (x, R)} \geq \frac{c}{V (x, f (x, l))},$$

where $c_3 F_2 \leq l \leq c_4 F_2$ and $y \in B \left(p, \frac{\delta R}{4} \right)$. Finally let $S \geq 2R$

$$\widetilde{p}_l^{B(p,S)} (x, y) \geq \widetilde{p}_l^{B(x,R)} (x, y) \geq \frac{c}{V (x, f (x, l))} \tag{9.10}$$

under the same conditions. Now choosing $\varepsilon' = \frac{c_F}{C_F} 2^{-\beta}$ and $\delta' = \frac{\delta}{4} (c_3 c_f)^{1/\beta'}$, (9.10) implies $PLE (F)$ (in the stronger form (9.4)):

$$\widetilde{p}_l^{B(p,S)} (x, y) \geq \frac{c}{V (x, f (x, l))} \tag{9.11}$$

for $d (x, y) \leq \delta' f (p, n), n \leq \varepsilon' F (p, S)$.

∎

9.2 Particular lower estimate

With this section the elliptic Harnack inequality enters the scene again and becomes a key player. The majority of the forthcoming results are based on the assumption that the elliptic Harnack inequality (H) holds on a weighted graph. This is a very interesting and essential condition, first it was shown by Moser [71] that it is satisfied in $\mathbb{R}^n$. Several works have been devoted to the study and use of the Harnack inequality, but stability against rough isometry (and other related questions) is still open (for history and recent progress see [4],[8],[70] and the bibliography there). In general, it is not easy to check the elliptic Harnack inequality in a space, still it is widely used and as our results show, in many cases it is necessary and sufficient.

9.2.1 Bounded oscillation

Now we start to develop the off-diagonal lower estimate. We follow the ideas of [48] to derive a near diagonal lower estimate $NDLE(F)$.

Definition 9.4. *On (Γ, μ) the near diagonal lower estimate $NDLE(F)$ holds if for all $x, y \in \Gamma, n > 0$ there are $c, \delta > 0$ such that*

$$\widetilde{p}_n(x, y) = p_n(x, y) + p_{n+1}(x, y) \geq \frac{c}{V(x, e(x, n))},$$

provided that $d(x, y) \leq \delta e(x, n)$.

The roadmap of the proof of the local lower estimate is as follows. We know that $(H) \Longrightarrow (MV)$ which means that from Theorem 8.2 we have

$$(VD) + (TC) + (H) \Longrightarrow DUE(E). \tag{9.12}$$

We also know that $(\overline{E}) \Longrightarrow DLE(E)$. In what comes, we show that (H) implies bounded oscillation of the heat kernel, and consequently, we have

$$(VD) + (TC) + DUE(E) + DLE(E) + (H) \Longrightarrow NDLE(E). \tag{9.13}$$

Definition 9.5. *For any set U and a function u on U, write*

$$\operatorname*{osc}_{U} u = \max_{U} u - \min_{U} u.$$

Proposition 9.1. *Assume that the elliptic Harnack inequality (H) holds on (Γ, μ). Then for any $\varepsilon > 0$, there exists $\sigma = \sigma(\varepsilon, H) < 1$ such that for any function u harmonic in $B(x, R)$, $R > 0$,*

$$\operatorname*{osc}_{B(x, \sigma R)} u \leq \varepsilon \operatorname*{osc}_{B(x, R)} u. \tag{9.14}$$

Proof   Let us first prove that

$$\underset{B(x,R/2)}{\mathrm{osc}}\, u \leq (1-\delta)\, \underset{B(x,R)}{\mathrm{osc}}\, u, \tag{9.15}$$

where $\delta = \delta(H) \in (0,1)$. Then (9.14) will follow by iterating (9.15) in a shrinking sequence of balls.

Let us set $v = u - \min_{B(x,R)} u$. By the Harnack inequality (H),

$$\max_{B(x,R/2)} v \leq H \min_{B(x,R/2)} v,$$

whence

$$\max_{B(x,R/2)} u - \min_{B(x,R)} u \leq H \left(\min_{B(x,R/2)} u - \min_{B(x,R)} u \right),$$

and

$$\underset{B(x,R/2)}{\mathrm{osc}}\, u \leq (H-1) \left(\min_{B(x,R/2)} u - \min_{B(x,R)} u \right).$$

Similarly, we have

$$\underset{B(x,R/2)}{\mathrm{osc}}\, u \leq (H-1) \left(\max_{B(x,R)} u - \max_{B(x,R/2)} u \right).$$

Summing up these two inequalities, we obtain

$$\underset{B(x,R/2)}{\mathrm{osc}}\, u \leq \frac{1}{2}(H-1) \left(\underset{B(x,R)}{\mathrm{osc}}\, u - \underset{B(x,R/2)}{\mathrm{osc}}\, u \right),$$

whence (9.15) follows by iteration.   ∎

Proposition 9.2. *Assume that the elliptic Harnack inequality (H) holds on (Γ, μ). Let $u \in c_0(B(x,R))$ satisfy the equation $\Delta u = f$ in $B(x,R)$. Then for any positive $r < R$,*

$$\underset{B(x,\sigma r)}{\mathrm{osc}}\, u \leq 2\left(\overline{E}(x,r) + \varepsilon \overline{E}(x,R) \right) \max|f|, \tag{9.16}$$

where σ and ε are the same as in Proposition 9.2.

Proof   For simplicity, write $B_r = B(x,r)$. By the definition of the Green function, we have

$$u(y) = -\sum_{z \in B_R} G^R(y,z) f(z),$$

whence we obtain

$$\max|u| \leq \overline{E}(x,R) \max|f|.$$

Let $v \in c_0(B_r)$ solve the Dirichlet problem $\Delta v = f$ in B_r (see Figure 9.2). In the same way, we have

$$\max|v| \leq \overline{E}(x,r) \max|f|.$$

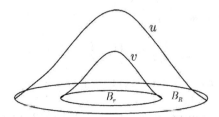

Fig. 9.2. The Dirichlet solution v

The function $w = u - v$ is harmonic in B_r whence by Proposition 9.2,

$$\operatorname*{osc}_{B_{\sigma r}} w \leq \varepsilon \operatorname*{osc}_{B_r} w.$$

Since $w = u$ on $\overline{B_r} \setminus B_r$, the maximum principle implies that

$$\operatorname*{osc}_{B_r} w = \operatorname*{osc}_{\overline{B_r} \setminus B_r} w = \operatorname*{osc}_{\overline{B_r} \setminus B_r} u \leq 2 \max |u|.$$

Hence,

$$\operatorname*{osc}_{B_{\sigma r}} u \leq \operatorname*{osc}_{B_{\sigma r}} v + \operatorname*{osc}_{B_{\sigma r}} w$$

$$\leq 2 \max |v| + 2\varepsilon \max |u|$$

$$\leq 2 \left(\overline{E}(x, r) + \overline{E}(x, R) \right) \max |f| ,$$

which was to be proved. ∎

9.2.2 Time derivative of the heat kernel

Definition 9.6. *Given a function $u_n(x)$ on $\Gamma \times \mathbb{N}$, by "the time derivative" of u we mean the difference*

$$D_n u = u_{n+2} - u_n .$$

Let us emphasize that this is *not* $u_{n+1} - u_n$.

Proposition 9.3. *Let A be a finite subset of Γ, f be a function on A and let us define*

$$u_n(x) = P_n^A f(x).$$

Then, for all integers $1 \leq k \leq n$,

$$\|D_n u\|_{L^2(A,\mu)} \leq \frac{1}{k} \|u_{n-k}\|_{L^2(A,\mu)} .$$

Proof   The proof follows an argument in [69]. Let $\phi_1, \phi_2, ..., \phi_{|A|}$ be the eigenfunctions of the Laplace operator $-\Delta_A$, and $\lambda_1, \lambda_2, ..., \lambda_{|A|}$ be the corresponding eigenvalues. Let us normalize ϕ_is to form an orthonormal basis in $L^2(A, \mu)$. On this basis the function f can be expanded:

$$f = \sum_i c_i \phi_i.$$

Since $P^A = I - (-\Delta_A)$, we obtain

$$u_n = \sum_i \rho_i^n \phi_i, \tag{9.17}$$

where $\rho_i = 1 - \lambda_i$ are eigenvalues of the transition matrix P^A.

In particular, we have

$$\|u_n\|_{L^2(A,\mu)}^2 = \sum_i \rho_i^{2n}.$$

From (9.17), we obtain

$$u_n - u_{n+2} = \sum_i \left(1 - \rho_i^2\right) \rho_i^n \phi_i$$

and

$$\|u_n - u_{n+2}\|_{L^2(A,\mu)}^2 = \sum_i \left(1 - \rho_i^2\right)^2 \rho_i^{2n}. \tag{9.18}$$

Note that $|\rho_i| \leq 1$, and hence $\rho_i^2 \in [0, 1]$. For any $a \in [0, 1]$, we have

$$1 \geq (1 + a + a^2 + ... + a^k)(1 - a) \geq k a^k (1 - a),$$

whence

$$(1 - a) a^k \leq \frac{1}{k}.$$

Applying this inequality to $a = \rho_i^2$, from (9.18) we obtain

$$\|u_n - u_{n+2}\|_{L^2(A,\mu)}^2 \leq \frac{1}{k^2} \sum_i \rho_i^{2(n-k)} = \frac{1}{k^2} \|u_{n-k}\|_{L^2(A,\mu)}^2,$$

which was to be proved.   ■

Proposition 9.4. *Let A be a finite subset of Γ. Then for all $x, y \in A$,*

$$\left| D_n p^A(x, y) \right| \leq \frac{1}{k} \sqrt{p_{2m}^A(x, x) p_{2(n-m-k)}^A(y, y)}, \tag{9.19}$$

for all positive integers n, m, k such that $m + k \leq n$.

Proof   From the semigroup identity applied to p^A, we obtain

$$D_n p^A(x,y) = \sum_{z \in A} p_m^A(x,z) D_{n-m} p^A(z,y) \mu(z),$$

whence

$$\left| D_n p^A(x,y) \right| \le \left\| p_m^A(x,\cdot) \right\|_{L^2(A,\mu)} \left\| D_{n-m} p^A(y,\cdot) \right\|_{L^2(A,\mu)}.$$

By Proposition 9.3,

$$\left\| D_{n-m} p^A(y,\cdot) \right\|_{L^2(A,\mu)} \le \frac{1}{k} \left\| p_{n-m-k}^A(y,\cdot) \right\|_{L^2(A,\mu)}$$

for any $1 \le k \le n - m$. Since

$$\left\| p_m^A(x,\cdot) \right\|_{L^2(A,\mu)}^2 = \sum_{z \in A} p_m^A(x,z)^2 \mu(z) = p_{2m}^A(x,x),$$

we obtain (9.19). ∎

Corollary 9.1. *If* $(VD), (TC)$ *and* $DUE(E)$ *hold, then there are* $C \ge 1$, $c > 0$ *such that for all* $x, y \in \Gamma$, $n \in \mathbb{N}$, $d(x,y) \le ce(x,n)$,

$$\left| D_n P^{B(x,R)}(x,y) \right| \le C \frac{\mu(y)}{n V(x, e(x,n))}. \tag{9.20}$$

Proof   If $n \le 3$, then (9.20) follows trivially from DUE because

$$|D_n p| \le p_n + p_{n+2}.$$

If $n > 3$, then choose k and m in (9.19) so that $k \sim m \sim n/3$ and $n - m - k \sim n/3$. Then

$$p_{2m}^A(x,x) \le \frac{C}{V(x, e(x, 2m))} \quad \text{and} \quad p_{2(n-m-k)}^A(y,y) \le \frac{C}{V(x, e(x, 2(n-m-k)))},$$

and (9.20) follows from (9.19) and from the regularity of V and e. ∎

9.2.3 Near diagonal lower estimate

From the previous propositions, the following particular lower estimate can be deduced.,

Proposition 9.5. *Assume* (p_0)*, then from* $DUE\,(E)$*,* $(\overline{E})$ *and* (H) *it follows that there are* $\delta, c, C > 0$ *such that for all* $x, y \in \Gamma, n \in \mathbb{N}, R \geq Ce(x, n)$*, the inequality* $PLE\,(E)$

$$p_n^{B(x,R)}(x,y) + p_{n+1}^{B(x,R)}(x,y) \geq \frac{c}{V(x, e(x, n))}, \tag{9.21}$$

and $NDLE\,(E)$

$$\widetilde{p}_n(x, y) = p_n(x, y) + p_{n+1}(x, y) \geq \frac{c}{V(x, e(x, n))}$$

hold, provided that $d(x, y) \leq \delta e(x, n)$*.*

Proof   Let us fix $x \in \Gamma$, $n \in \mathbb{N}$ and set

$$R = e\,(x, 2n)\,. \tag{9.22}$$

Write $A = B(x, R)$ and introduce the function

$$\widetilde{u}(y) = p_{2n}^A(x, y) + p_{2n+1}^A(x, y).$$

From the assumption $(\overline{E})$ it follows by Theorem 6.2 that we have $DLE\,(E)$ which means that

$$\widetilde{u}(x) \geq cV(x, e(x, 2n))^{-1}.$$

Let us show that

$$|\widetilde{u}(x) - \widetilde{u}(y)| \leq \frac{c}{2} \frac{1}{V(x, e(x, 2n))}, \tag{9.23}$$

provided that $d(x, y) \leq \delta e(x, 2n)$, which would imply $\widetilde{u}(y) \geq \frac{c}{2} V(x, e(x, 2n))^{-1}$, and hence prove (9.21).

The function $u(y)$ is in class $c_0(A)$ and solves the equation $\Delta u(y) = f(y)$, where

$$f(y) = p_{2n+2}^A(x, y) - p_{2n}^A(x, y).$$

The on-diagonal upper bound $DUE\,(E)$ implies, by Corollary 9.1,

$$\max |f| \leq \frac{C}{nV(x, e(x, 2n))}. \tag{9.24}$$

By (H) and Proposition 9.2, for any $0 < r < R$ and for some $\sigma = \sigma(\varepsilon^2) \in (0, 1)$, we have

$$\operatorname*{osc}_{B(x,\sigma r)} \widetilde{u} \leq 2 \left(\overline{E}(x, r) + \varepsilon^2 \overline{E}(x, R) \right) \max |f|\,.$$

From the assumption $(\overline{E})$ we can estimate $\overline{E}$ with E. Estimating $\max |f|$ by (9.24), we obtain

$$\operatorname*{osc}_{B(x,\sigma r)} u \leq C \frac{E(x, r) + \varepsilon^2 E(x, R)}{V(x, e(x, 2n))}.$$

Choosing r to satisfy $r = \varepsilon R$ and substituting $n = \varepsilon E(x, R)$ from (9.22), we obtain

$$\underset{B(x,\sigma r)}{\mathrm{osc}}\, \widetilde{u} \leq C \frac{\varepsilon^2 E(R)}{2nV(x, e(x, n))} = \frac{\varepsilon C}{V(x, e(x, 2n))}.$$

From this it follows that

$$\underset{B(x,\sigma r)}{\mathrm{osc}}\, \widetilde{u} \leq \frac{c}{2} \frac{1}{V(x, e(x, 2n))}, \tag{9.25}$$

provided that ε is small enough which implies

$$|\widetilde{u}(x) - \widetilde{u}(y)| \leq \frac{c}{2} \frac{1}{V(x, e(x, 2n))},$$

if $d(x, y) \leq \delta e(x, 2n)$. ∎

Proposition 9.6. *For weighted graphs*

$$(p_0) + (VD) + (TC) + (H) \Longrightarrow NDLE\,(E).$$

Proof   One should note that (H) implies (MV), hence also $DUE\,(E)$ due to Theorem 8.2. We know that (VD) and (TC) imply $(\overline{E})$, and from that the diagonal lower estimate DLE follows as well which means that all the conditions of Proposition 9.5 are satisfied and hence $NDLE$ follows. ∎

9.3 Lower estimates without upper ones

This is a special section. The reader can skip it and jump to the next one if she or he is more curious to know how two-sided heat kernel estimates are established in the local theory. This chapter is a detour towards to the problem indicated in the title above. Typical proofs of the lower estimate (as it was given in Section 9.2.3) are based on the diagonal upper estimate and use assumptions on the volume. Here a different approach is taken which demonstrates, among others things, the power of the elliptic Harnack inequality.

We have seen that (TC) or even $(\overline{E})$ implies

$$P\,(T_{x,R} < n) \leq C \exp\left[-ck_x\,(n, R)\right]$$

One might wonder which condition ensures (up to the constant) the same lower bound. The answer is given with the aid of a new chaining argument.

Definition 9.7. *For $x, y \in \Gamma, n \geq R > 0, C > 0$, let us define $l = l_C\,(x, y, n, R)$ as the minimal integer for which*

$$\frac{n}{l} \geq Q \max_{z \in \pi_{x,y}} E\left(z, \frac{CR}{l}\right),$$

where Q is a fixed constant (to be specified later). Let $l = R$ by definition if there is no such integer. If $d\,(x, y) = R$, we will use the shorter notation $l_C\,(x, y, n) = l_C\,(x, y, n, d\,(x, y))$.

Definition 9.8. *For a given* $x \in \Gamma, n \geq R > 0$ *let us define*

$$\nu = \nu(x, n, R) = \min_{y \in B(x,R)^c} l_9(x, y, n, R).$$

Theorem 9.2. *Assume that a weighted graph* (Γ, μ) *satisfies* (p_0).
1. If $(\overline{E})$ *holds , then there are* $c, C > 0$ *such that for all* $n \geq R > 0.x \in \Gamma$

$$P(T_{x,R} < n) \leq C \exp[-ck_x(n, R)]$$

is true.

2. If (Γ, μ) *satisfies the elliptic Harnack inequality* (H), *then there are* $c, C > 0$ *such that* $n \geq R > 0.x \in \Gamma$

$$P(T_{x,R} < n) \geq c \exp[-C\nu(x, n, R)]. \tag{9.1}$$

The lower bound is based on a chaining argument. First we need some propositions.

Lemma 9.2. *If* (Γ, μ) *satisfies* (p_0) *and the elliptic Harnack inequality* (H), *then there is a* $c_1 > 0$ *such that for all* $x \in \Gamma, r > 0$ *and* $w \in B(x, 4r)$,

$$\mathbb{P}_w(\tau_{x,r} < T_{x,5r}) > c_1. \tag{9.2}$$

Proof   The investigated probability

$$u(w) = \mathbb{P}_w(\tau_{x,r} < T_{x,5r}) \tag{9.3}$$

is the capacity potential between $B^c(x, 5r)$ and $B(x, r)$ and clearly harmonic in $A = B(x, 5r) \setminus B(x, r)$. So it can be as usual decomposed

$$u(w) = \sum_z g^{B(x,5r)}(w, z) \pi(z)$$

with the proper capacity measure $\pi(z)$ with support in $S(x, r)$, $\pi(A) = 1/\rho(x, r, 5r)$. From the maximum (minimum) principle it follows that the minimum of $u(w)$ is attained on the boundary, $w \in S(x, 4r - 1)$ and from the Harnack inequality for $g^{B(x,5r)}(w, .)$ in $B(x, 2r)$ that

$$\min_{z \in \overline{B}(x,r+1)} g^{B(x,5r)}(w, z) \geq cg^{B(x,5r)}(w, x),$$

$$u(w) = \sum_z g^{B(x,5r)}(w, z) \pi(z) \geq \frac{cg^{B(x,5r)}(w, x)}{\rho(x, r, 5r)}.$$

From Proposition 3.7 we know that

$$\max_{y \in B(x,5r) \setminus B(x,4r)} g^{B(x,5r)}(y, x) \simeq \rho(x, 4r, 5r) \simeq \min_{w \in B(x,4r)} g^{B(x,5r)}(w, x).$$

which means that

$$u\left(w\right) \geq c\frac{\rho\left(x, 4r, 5r\right)}{\rho\left(x, r, 5r\right)}.\tag{9.4}$$

Similarly from Proposition 3.7 it follows that

$$\max_{y \in B(x,5r) \setminus B(x,r)} g^{B(x,5r)}\left(v, x\right) \simeq \rho\left(x, r, 5r\right) \simeq \min_{w \subset B(x,r)} g^{B(x,5r)}\left(w, x\right).$$

Finally if y_0 is on the ray from x to y then iterating the Harnack inequality along a finite chain of balls of radius $r/4$ along this ray from y_0 to y one obtains

$$g^{B(x,5r)}\left(y, x\right) \simeq g^{B(x,5r)}\left(y_0, x\right),$$

which results that

$$\rho\left(x, 4r, 5r\right) \geq c\rho\left(x, r, 5r\right),$$

and the statement follows from (9.4).   ∎

Proposition 9.7. *Assume that a weighted graph (Γ, μ) satisfies (p_0) and (H). Then there are $c_0, c_1 > 0$ such that for all $x, y \in \Gamma, r > 0, d\left(x, y\right) < 4r, m > \frac{2}{c_1} E\left(x, 9r\right)$*

$$\mathbb{P}_x\left(\tau_{y,r} < m\right) > c_0.$$

Proof   We start with the following simple estimate:

$$\begin{aligned}
\mathbb{P}_x\left(\tau_{z,r} < m\right) &\geq \mathbb{P}_x\left(\tau_{z,r} < T_{x,9r} < m\right)\\
&= \mathbb{P}_x\left(\tau_{z,r} < T_{x,9r}\right) - \mathbb{P}_x\left(\tau_{z,r} < T_{x,9r}, T_{x,9r} \geq m\right)\\
&\geq \mathbb{P}_x\left(\tau_{z,r} < T_{x,9r}\right) - \mathbb{P}_x\left(T_{x,9r} \geq m\right).
\end{aligned}$$

On the one hand,

$$\mathbb{P}_x\left(T_{x,9r} \geq m\right) \leq \frac{E\left(x, 9r\right)}{m} \leq \frac{E\left(x, 9r\right)}{\frac{2}{c_1} E\left(x, 9r\right)} < c_1/2,$$

on the other hand $B\left(z, 5r\right) \subset B\left(x, 9r\right)$, hence

$$\mathbb{P}_x\left(\tau_{z,r} < T_{x,9r}\right) \geq \mathbb{P}_x\left(\tau_{z,r} < T_{z,5r}\right),$$

and Lemma 9.2 can be applied to get

$$\mathbb{P}_x\left(\tau_{z,r} < T_{z,5r}\right) \geq c_1.$$

The result follows by choosing $c_0 = c_1/2$.   ∎

Lemma 9.3. *Let us assume that* $x \in \Gamma, m, r, l \geq 1$, *write* $n = ml, 0 \leq u \leq 3l$, $R = (3l - 2)r - u$, $y \in S(x, R + r)$, *then*

$$P_x(\tau_{y,r} < n) \geq \min_{w \in \pi_{x,y}, 2r - 3 \leq d(z,w) \leq 4r} \mathbb{P}_z^l(\tau_{w,r} < m).$$

where $\pi_{x,y}$ *is the union of vertices of the shortest paths from* x *to* y.

Proof   We define a chain of balls. For $1 \leq l \leq d(x, y) - r$ let us consider a sequence of vertices $x_0 = x, x_1, ...x_l = y, x_i \in \pi_{x,y}$ in the following way: $d(x_{i-1}, x_i) = r - \delta_i$, where $\delta_i \in \{0, 1, 2, 3\}$ for $i = 1...l$ and

$$u = \sum_{i=1}^{l} \delta_i$$

$$R = (3l - 2)r - \sum_{i=1}^{l} \delta_i = (3l - 2)r - u.$$

Let $\tau_i = \tau_{x_i,r}$ and $s_i = \tau_i - \tau_{i-1}$, $A_i = \{s_i < m\}$ for $i = 1, ...l$, $\tau_0 = 0$. Let us use

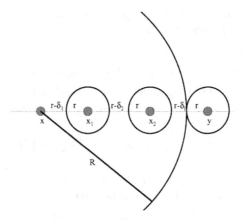

Fig. 9.3. Chain of balls

the notation $D_i(z_i) = A_i \cap \{X_{\tau_i} = z_i\}$. One can observe that $\cap_{i=1}^{l} A_i$ means that the walk takes less than m steps between the first hit of the consecutive $B_i = B(x_i, r)$ balls, consequently

$$\mathbb{P}_x(\tau_{y,r} < n) \geq \mathbb{E}_x\left(\cap_{i=1}^{l} A_i\right)$$

We also note that $s_i = \min\{k : X_k \in B_k | X_0 \in \partial B_{i-1}\}$. From this one obtains the following estimates denoting $z_0 = x$

$$\mathbb{P}_x\left(\tau_{y,r} < n\right) \geq \tag{9.5}$$

$$\mathbb{P}_x\left(\cap_{i=1}^l A_i\right) = \mathbb{E}_x\left(\cap_{i=1}^l \cup_{z_i \in \partial B_i} D_i\left(z_i\right)\right)$$

$$= \mathbb{E}_x\left(\cup_{z_1 \in \partial B_1} \cup_{z_2 \in \partial B_2} \cdots \cup_{z_l \in \partial B_l} \cap_{i=1}^l D_i\left(z_i\right)\right).$$

$$= \sum_{z_1 \in \partial B_1} \sum_{z_2 \in \partial B_2} \cdots \sum_{z_l \in \partial B_l} \mathbb{P}_x\left[\cap_{i=1}^l D_i\left(z_i\right)\right]$$

Now we use the Markov property.

$$\sum_{z_1 \in \partial B_1} \sum_{z_2 \in \partial B_2} \cdots \sum_{z_l \in \partial B_l} \mathbb{P}_x\left[D_i\left(z_l\right) \mid \cap_{i=1}^{l-1} D_i\left(z_i\right)\right] \mathbb{P}_x\left[\cap_{i=1}^{l-1} D_i\left(z_i\right)\right] \tag{9.6}$$

$$\sum_{z_1 \in \partial B_1} \sum_{z_2 \in \partial B_2} \cdots \sum_{z_l \in \partial B_l} \mathbb{P}_{z_{l-1}}\left(s_l < m, X_{\tau_l} = z_l\right) \mathbb{P}\left(\cap_{i=1}^{l-1} D_i\left(z_i\right)\right)$$

$$= \sum_{z_1 \in \partial B_1} \sum_{z_2 \in \partial B_2} \cdots \sum_{z_{l-1} \in \partial B_{l-1}} \mathbb{P}_{z_{l-1}}\left(s_l < m\right) \mathbb{P}\left(\cap_{i=1}^{l-1} D_i\left(z_i\right)\right)$$

$$= \ldots = \min_{w \in \pi_{x,y}, 2r-3 \leq d(z,w) \leq 4r} \mathbb{P}_z^l\left(\tau_{w,r} < m\right).$$

∎

Proposition 9.8. *Assume that a weighted graph (Γ, μ) satisfies (p_0) and the elliptic Harnack inequality (H). Then there are $c, C > 0$ such that for all $x, y \in \Gamma, r \geq 1, n > d(x,y) - r$*

$$\mathbb{P}_x\left(\tau_{y,r} < n\right) \geq c \exp\left[-Cl_9\left(x,y,n,d(x,y) - r\right)\right].$$

Proof   If $n > \frac{2}{c_1} E(x, 9R)$, then the statement follows from Proposition 9.7. Also, if $r \leq 9$, then $\frac{R}{3r} \leq l \leq R$, so from (p_0) the trivial lower estimate

$$\mathbb{P}_x\left(\tau_{y,r} < n\right) \geq c \exp\left[-27\left(\log\frac{1}{p_0}\right) l\right]$$

gives the statement. If $n < \frac{2}{c_1} E(x, 9R)$ and $r \geq 10$, then $l_9(x,y,n,R) > 1$ and $R = (3l - 2)r - u \geq 34$. Let us use Lemma 9.3:

$$\mathbb{P}_x\left(\tau_{y,r} < n\right) \geq \min_{w \in \pi_{x,y}, 2r-3 \leq d(z,w) \leq 4r} \mathbb{P}_z^l\left(\tau_{w,r} < m\right),$$

and by Proposition 9.7 $\mathbb{P}_z^l\left(\tau_{w,r} < m\right) > c$ if for $w \in \pi_{x,y}$

$$\frac{n}{l} > \frac{2}{c_1} E(w, 9r) = \frac{2}{c_1} E\left(w, 9\frac{R+u}{3l-2}\right) \tag{9.7}$$

is satisfied. Consider the following straightforward estimates for $r \geq 10, R \geq 10$.

$$9r = 10(r-1) \leq 10\left(\frac{R+u}{3l-2} - 1\right) \leq 10\left(\frac{R+3l}{3l-2} - 1\right) = 10\frac{R+2}{3l-2}$$

$$\leq \frac{4R}{(l-1)} \leq 8\frac{R}{l} < 9\frac{R}{l}.$$

If $l = l_9(x, y, n, R)$, then the inequality (9.7) is satisfied, and Proposition 9.7 can be applied to get a uniform lower bound for all $\mathbb{P}^l_z(\tau_{w,r} < m)$. ∎

Proof [of Theorem 9.2] The lower bound is immediate from Proposition 9.8 minimizing $l_9(x, y, n)$ for $d = d(x, y) = 2R, y \in S(x, 2R), \frac{d}{4} = R/2 \le r < R$

$$\mathbb{P}_x(T_{x,R} < n) \ge \mathbb{P}_x(\tau_{y,r} < n).$$

∎

9.3.1 Very strongly recurrent graphs

Definition 9.9. *Using the terminology of [3], we say that a graph is very strongly recurrent (VSR) if there is a $c > 0$ such that for all $x \in \Gamma, r > 0, w \in \partial B(x, r)$,*

$$\mathbb{P}_w(\tau_x < T_{x,2r}) \ge c.$$

In this section we deduce an off-diagonal heat kernel lower bound for very strongly recurrent graphs. The proof is based on Theorem 9.2 and the fact that very strong recurrence implies the elliptic Harnack inequality (cf. [3]). Let we mention here strong recurrence defined in [98] (cf. 8.1) and it can be seen easily that strong recurrence in conjunction with the elliptic Harnack inequality is equivalent to very strong recurrence. It is worth noting that typically considered finitely ramified fractals and their pre-fractal graphs are (very) strongly recurrent.

Theorem 9.3. *Let us assume that (Γ, w) satisfies (p_0) and is very strongly recurrent, furthermore it satisfies $(\overline{E})$. Then there are $C, c > 0$ such that for all $x, y \in \Gamma, n > 0$,*

$$\widetilde{p}_n(x, y) \ge \frac{c}{V(x, e(x, n))} \exp\left[-Cl_9\left(x, y, \frac{1}{2}n, d\right)\right], \tag{9.8}$$

where $d = d(x, y)$.

Remark 9.2. Typical examples for very strongly recurrent graphs are pre-fractal skeletons of p.c.f. self-similar sets (for definition and further reading see [3] and [4]). We recall Barlow's [3] and Delmotte's [30] constructions. Let us consider two trees Γ_1, Γ_2 which are (VSR) and assume that $V_i(x, R) \simeq R^{\alpha_i}, E(x, R) \simeq R^{\beta_i}, \alpha_1 \ne \alpha_2,$

$$\gamma = \beta_1 - \alpha_1 = \beta_2 - \alpha_2 > 0,$$

which basically means that

$$\rho(x, R, 2R) \simeq R^\gamma$$

for both graphs. Such trees are constructed in [3]. Let Γ be the joint of Γ_1 and Γ_2 which means that two vertices O_1, O_2 are chosen and identified (for details see [31]). The resulting graph satisfies the Harnack inequality but not the volume doubling property. Using the fact that Γ_i are trees, we can also see that Γ is a (VSR) tree as well. This means that Γ is an example of graphs that satisfy the Harnack inequality but not the usual volume properties.

It was realized some time ago that the so-called near diagonal lower bound (9.9) is a crucial step to obtain off-diagonal lower estimates. Here we utilize the fact that the near diagonal lower bound is an easy consequence of very strong recurrence.

Proposition 9.9. *Let us assume that (Γ, μ) satisfies (p_0). If the graph is very strongly recurrent and $(\overline{E})$ holds, then there are $c, c' > 0$ such that for all $x, y \in \Gamma, m \geq \frac{2}{c'} E(x, 2d(x,y))$,*

$$\widetilde{p}_m(x,y) \geq \frac{c}{V(x, e(x, m))}. \tag{9.9}$$

Proof   The proof starts with a first hit decomposition and uses Theorem 6.2, in particular DLE.

$$\widetilde{p}_m(y,x) \geq \sum_{i=0}^{m-1} \mathbb{P}_y(\tau_x = i)\, \widetilde{p}_{m-i}(x,x) \geq \mathbb{P}_y(\tau_x < m)\, \widetilde{p}_m(x,x)$$

$$\geq \frac{c}{V(x, e(x, m))} \mathbb{P}_y(\tau_x < m).$$

We estimate the latter term as in the proof of Proposition 9.7. Write $r = d(x,y)$,

$$\mathbb{P}_y(\tau_x < m) \geq \mathbb{P}_y(\tau_x < T_{x,2r} < m) \geq \mathbb{P}_y(\tau_x < T_{x,2r}) - \mathbb{P}_y(T_{x,2r} \geq m).$$

From (VSR) we have that $\mathbb{P}_y(\tau_x < T_{x,2r}) > c$, so from $m \geq \frac{2}{c'} E(x, 2r)$ and from the Markov inequality it follows that

$$\mathbb{P}_y(T_{x,2r} \geq m) \leq \frac{E(x, 2r)}{m} \leq c'/2.$$

Consequently we have that $\mathbb{P}_y(\tau_x < s) > c'/2$ and we obtain the result.   ∎

Proof   [of Theorem 9.3] If $l = l_9(x, y, n, d(x, y)) = 1$, then $n > \frac{2}{c'} E(x, 9d) >$

$\frac{2}{c'} E(x, 2d)$, and the statement follows from Proposition 9.9. Let us assume that $l > 1$ and start with a path decomposition. Write $m = \lfloor \frac{n}{l} \rfloor$, $r = \lfloor \frac{R}{l} \rfloor$, $S = \{y : d(x,y) = r\}$ and $\tau = \tau_S$,

$$\widetilde{p}_n\left(y,x\right) = \frac{1}{\mu\left(x\right)}\mathbb{P}_y\left(X_n = x \text{ or } X_{n+1} = x\right)$$

$$\geq \sum_{i=0}^{n-m-1}\sum_{w\in S}\mathbb{P}_y\left(X_\tau = w, \tau = i\right)\min_{w\in S}\widetilde{p}_{n-i}\left(w,x\right)$$

$$\geq \sum_{i=0}^{n-m-1}\mathbb{P}_y\left(\tau = i\right)\min_{w\in S}\widetilde{p}_{n-i}\left(w,x\right).$$

The next step is to use the near diagonal lower estimate:

$$\widetilde{p}_n\left(y,x\right) \geq \sum_{i=0}^{n-m-1}\mathbb{P}_y\left(\tau = i\right)\min_{w\in S}\widetilde{p}_{n-i}\left(w,x\right)$$

$$\geq \sum_{i=0}^{n-m-1}\mathbb{P}_y\left(\tau = i\right)\frac{c}{V\left(x,e\left(x,n-i\right)\right)}$$

$$\geq \mathbb{P}_y\left(\tau < \frac{n}{2}\right)\frac{c}{V\left(x,e\left(x,n\right)\right)}.$$

In the proof of Theorem 9.2 we have seen that

$$\mathbb{P}_y\left(\tau_{x,r} < \frac{n}{2}\right) \geq c\exp\left[-Cl_9\left(x,y,\left(\frac{n}{2}\right),d-r\right)\right],$$

which finally yields

$$\widetilde{p}_n\left(y,x\right) \geq \frac{c}{V\left(x,e\left(x,n\right)\right)}\exp\left[-Cl_9\left(x,y,\frac{n}{2},d-r\right)\right]$$

$$\geq \frac{c}{V\left(x,e\left(x,n\right)\right)}\exp\left[-Cl_9\left(x,y,\frac{1}{2}n,d\right)\right].$$

∎

9.3.2 Harnack inequality implies a lower bound

In this section the following off-diagonal lower bound is proved.

Theorem 9.4. *Let us assume that the graph $\left(\Gamma,\mu\right)$ satisfies $\left(p_0\right)$. We also suppose that the time comparison principle (2.14) and the elliptic Harnack inequality $\left(H\right)$ hold. Then there are $c, C, D > 0$ constants such that for any $x, y \in \Gamma, n \geq d\left(x,y\right),$*

$$\widetilde{p}_n\left(x,y\right) \geq \frac{c}{V\left(x,e\left(x,n\right)\right)r^D}\exp\left[-Cl_9\left(x,y,\frac{n}{2}\right)\right], \tag{9.10}$$

where $R = d\left(x,y\right), r = \frac{R}{3l}, l = l_9\left(x,y,\frac{n}{2}\right).$
In particular, if $n < c\frac{E\left(x,R\right)}{\left(\log E\left(x,R\right)\right)^{\beta-1}},$ then

$$\widetilde{p}_n\left(x,y\right) \geq \frac{c}{V\left(x,e\left(x,n\right)\right)}\exp\left[-Cl_9\left(x,y,\frac{n}{2}\right)\right]. \tag{9.11}$$

Corollary 9.2. *If we assume* (p_0), $E \in V_1$ *and that the elliptic Harnack inequality* (H) *is true, then*

$$\tilde{p}_n(x,y) \geq \frac{c}{V(x,e(x,n))} \exp\left(-C\left[\frac{E(x,R)}{n}\right]^{\frac{1}{\beta'-1}}\right)$$

for $n < c\frac{E(x,R)}{(\log E(x,R))^{\beta'-1}}$, *where* $R = d(x,y)$.

This corollary is a direct consequence of Theorem 9.4.

Proposition 9.10. *Let us assume that* $E \in V_0$ *and that the Harnack inequality* (H) *holds. Then there are* $D, c > 0$ *such that for* $x, y \in \Gamma$, $r = d(x,y)$, $m > CE(x,r)$ *the inequality*

$$\tilde{p}_m(y,x) \geq \frac{c}{V(x,e(x,m))} r^{-D}$$

holds.

Proof   The proof is based on a modified version of the chaining argument used in the proof of Lemma 9.3. From Proposition 9.9 we know that $(\overline{E})$ implies

$$\tilde{p}_n(x,x) \geq \frac{c}{V(x,e(x,n))}. \tag{9.12}$$

Let us recall (7.5) and set $A = \max\{9, A_T\}$, $K = \lceil \frac{4}{4} \rceil$. Consider a sequence of times $m_i = \frac{m}{2^i}$ and radii $r_i = \frac{r}{A^i}$. From the condition $m > CE(x,r)$ and (7.5) it follows that for all i

$$m_i > CE(x,r_i) \tag{9.13}$$

holds as well. Let us write $B_i = B(x,r_i)$, $\tau_i = \tau_{B_i}$ and start a chaining.

$$\tilde{p}_m(y,x) = \sum_{k=1}^{m} P(\tau_1 = k) \min_{w \in \partial B_1} \tilde{p}_{m-k}(w,x) \geq \sum_{i=1}^{m/2} P(\tau_1 = k) \min_{w \in \partial B_1} \tilde{p}_{m-k}(w,x)$$

$$\geq P(\tau_1 < m/2) \min_{1 \leq k \leq m/2} \min_{w \in \partial B_1} \tilde{p}_{m-k}(w,x).$$

Let us continue in the same way for all $i \leq L = \lceil \log_A r \rceil$ (see Figure 9.4). It is clear that $B_L = \{x\}$ which yields

$$\tilde{p}_m(y,x) \geq \min_{w_i \in \partial B_i} P_y(\tau_1 < m/2) \dots$$

$$\dots P_{w_j}\left(\tau_j < \frac{m}{2^i}\right) \dots P\left(\tau_L < \frac{m}{2^L}\right) \min_{0 \leq k \leq m-L} \tilde{p}_k(x,x).$$

From the initial conditions and (7.5) we have (9.13) for all j, so we can use a slight modification of Proposition 9.7 to get

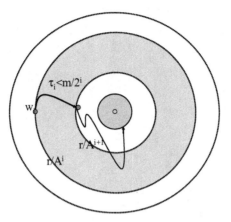

Fig. 9.4. Chaining inwards

$$P_w\left(\tau_i < \frac{m}{2^i}\right) > c_2$$

for all $w_j \in B\left(x, r_j\right)$ and j. Consequently, using (9.12) we have

$$\widetilde{p}_m\left(y, x\right) \geq \frac{c}{V\left(x, e\left(x, m\right)\right)} c_2^L \geq \frac{c}{V\left(x, e\left(x, m\right)\right)} r^{-D}$$

where $D = \frac{\log \frac{1}{c_2}}{\log A}$. ■

Proof [of Theorem 9.4] The proof is a combination of two chainings. Let us recall that the time comparison principle implies (7.5), so the conditions of Proposition 9.10 are satisfied. First let us use Theorem 9.2 to reach the boundary of $B\left(x, r\right)$, where $r = \frac{d(x,y)}{3l-1}, l = l_9\left(x, y, \frac{n}{2}, d\left(x, y\right)\right)$, then let us use Proposition 9.10. ■

10

Two-sided estimates

In this chapter we harvest the crop of the earlier chapters. In order to show the conjunction of the upper and lower estimates, we use key observations on the Einstein relation, on the elliptic Harnack inequality and on upper and lower heat kernel estimates.

Corollary 10.1. *Assume that (Γ, μ) satisfies $(p_0), (VD)$ and (H), then*

$$(wTC) \Leftrightarrow (aD\rho v) \Leftrightarrow (TC) \Leftrightarrow (ER) \Leftrightarrow RLE(E) \overset{(i)}{\Leftrightarrow} g(F), \qquad (10.1)$$

where $F \in W_0$ is a consequence in the direction $\overset{(i)}{\Longrightarrow}$ and an assumption for $\overset{(i)}{\Longleftarrow}$.

Proof All the implications are established in Remark 7.2, except the last one which is given there for E, i.e., for $g(E)$ and $RLE(E)$. The implication $(ER) \Longrightarrow RLE(E)$ is trivial. Its reverse is similarly simple; $c\rho v \leq E$ follows as usual, and $E < C\rho v$ is $RLE(E)$ itself. It is clear that the $\overset{(i)}{\Longrightarrow}$ direction holds and only its reverse needs some additional arguments. Let us assume that $r_i = 2^i, r_{n-1} < 2R \leq r_n, B_i = B(x, r_i), A_i = B_i \backslash B_{i-1}$ and $V_i = V(x, r_i)$. We have derived in Proposition 7.5 from $(p_0), (VD)$ and (H) that

$$E(x, 2R) \leq C \sum_{i=0}^{n-1} V_{i+1}\rho(x, r_i, r_{i+1}).$$

Now we use a consequence (3.39) of (H):

$$\rho(x, r_i, r_{i+1}) \leq C \max_{y \in A_{i+1}} g^{B_{i+1}}(x, y)$$

to obtain

$$E\left(x,2R\right) \leq C \sum_{i=0}^{n-1} V_{i+1} \max_{y \in A_{i+1}} g^{B_{i+1}}\left(x,y\right)$$

$$\leq C \sum_{i=0}^{n-1} F\left(x,r_{i+1}\right) \leq CF\left(x,r_n\right) \sum_{i=0}^{n-1} 2^{-i\beta'}$$

$$\leq CF\left(x,2R\right),$$

where (7.12) was used to get the second inequality.

On the other hand, from (7.13) we obtain

$$c \frac{F\left(x,2R\right)}{V\left(x,2R\right)} \left(V\left(x,R\right) - V\left(x,R/2\right)\right)$$

$$\leq \min_{y \in B(x,R) \setminus B(x,R/2)} g^{B}\left(x,y\right) \sum_{z \in B(x,R) \setminus B(x,R/2)} \mu\left(z\right)$$

$$\leq \sum_{z \in B(x,R) \setminus B(x,R/2)} g^{B}\left(x,z\right) \mu\left(z\right) \leq E\left(x,2R\right),$$

which means that

$$cF\left(x,2R\right) \leq E\left(x,2R\right).$$

Consequently, $F \simeq E$, $E \in W_0$ and (TC) is satisfied which implies (ER) and all the other equivalent conditions. ∎

Remark 10.1. Let us remark here that as a side result it follows that $RLE\left(E\right)$ or for $F \in W_0$, $g\left(F\right)$ implies $\rho v \simeq F$ and $E \simeq F$ as well.

We have seen that

$$\left(wTC\right) \Leftrightarrow \left(aD\rho v\right) \Leftrightarrow \left(TC\right) \Leftrightarrow \left(ER\right) \Leftrightarrow RLE\left(E\right). \tag{10.2}$$

Let $(*)$ denote any of the equivalent conditions. Using this convention, we can state the main result on weakly homogeneous graphs.

Theorem 10.1. *If a weighted graph (Γ,μ) satisfies (p_0) and (VD), then the following statements are equivalent:*

1. *there is an $F \in W_0$ such that $g\left(F\right)$ is satisfied;*
2. *(H) and (wTC) hold;*
3. *(H) and (TC) hold;*
4. *(H) and $(aD\rho v)$ hold;*
5. *(H) and (ER) hold;*
6. *(H) and $RLE\left(E\right)$ hold;*
7. *there is an $F \in W_0$ such that $UE\left(F\right)$ and $PLE\left(F\right)$ are satisfied;*
8. *there is an $F \in W_0$ such that $PMV\left(F\right)$ and $PSMV\left(F\right)$ are satisfied.*

The proof of Theorem 10.1 contains two autonomous results. The first one is $UE\,(F) \Longleftrightarrow PMV\,(F)$, the second one is $PLE\,(F) \Longleftrightarrow PSMV\,(F)$.

Let us emphasize the importance of the condition $(aD\rho v)$. It entirely relies on volume and resistance properties, no assumption of stochastic nature is involved, so the result is in the spirit of Einstein's observation on heat propagation. This condition in conjunction with (VD) and (H) provides the characterization of heat kernel estimates in terms of volume and resistance properties. Of course, the elliptic Harnack inequality is not easy to verify. We learn from $g\,(F)$ that the main properties ensured by the elliptic Harnack inequality that the equipotential surfaces of the local Green kernel $g^{B(x,R)}$ are basically spherical and that the potential growth is regular.

10.1 Time comparison (the return route)

In this section we summarize the results which lead to the equivalence of 1...6, and $6 \Longrightarrow 7$ in Theorem 10.1, and we prove the return route from $8 \Longrightarrow 2$. The equivalence of 1 and 6 is established by Theorem 7.1, 7.4 and 10.1, see also Remark 10.1. The implication $6 \Longrightarrow 7$ is given by Theorem 8.2, and $7 \Longleftrightarrow 8$ is a combination of Theorem 8.2 and 9.1.

Now we prove the return route $8 \Longrightarrow 2$ of Theorem 10.1.

Our task is to verify the implications in the diagram below under the assumption $F \in W_0,(p_0)$ and (VD).

$$\left.\begin{array}{c} PMV_1\,(F) \\ PSMV\,(F) \end{array}\right\} \Longrightarrow \left.\begin{array}{c} PMV_{\delta*}\,(F) \\ PSMV\,(F) \end{array}\right\} \Longrightarrow (H) \qquad (10.3)$$

$$\left.\begin{array}{c} PMV_1\,(F) \\ PSMV\,(F) \end{array}\right\} \Longrightarrow \left.\begin{array}{c} DUE\,(F) \\ PLE\,(F) \\ (H) \end{array}\right\} \Longrightarrow \begin{array}{c} \rho v \simeq F \\ (H) \end{array} \Longrightarrow \begin{array}{c} (TC) \\ (H) \end{array} \qquad (10.4)$$

The heat kernel estimates are established as we indicated above. Now we deal with the proof of the elliptic Harnack inequality (H) and the time comparison principle (TC).

Theorem 10.2. *If Γ satisfies $(p_0),(VD)$ and there is an $F \in W_0$ for which $PMV\,(F)$ and $PSMV\,(F)$ are satisfied, then the elliptic Harnack inequality holds.*

Lemma 10.1. *If (Γ,μ) satisfies $(p_0),(VD)$ and $PMV_1\,(F)$ for an $F \in W_0$, then for a given $\varepsilon, \delta > 0, 0 < \delta^* \leq \frac{1}{C_F}\varepsilon^{\frac{1}{\beta'}}\frac{\delta}{2}$ there are $c_1 < ... < c_4$ such that $PMV_{\delta*}\,(F)$ holds for ε and $c_i s$.*

Proof   We would like to derive $PMV_{\delta*}\,(F)$ for c_i from $PMV_1\,(F)$ which holds for some other constants a_i. We will apply $PMV_1\,(F)$ on the ball $B = B\,(x,\delta R)$ and re-scale the time accordingly. We have $PMV_{\delta*}\,(F)$ on $B\,(x,R)$ by

$$\max_{\substack{c_3 F(x,R)\le i\le c_4 F(x,R)\\ y\in B}} u_i \le \max_{\substack{a_3 F(x,\delta R)\le i\le a_4 F(x,\delta R)\\ y\in B}} u_i$$

$$\le \sum_{j=a_1 F(x,\delta R)}^{a_2 F(x,\delta R)} \sum_{y\in B} u_j(z)\,\mu(z) \le \sum_{j=c_1 F(x,R)}^{c_2 F(x,R)} \sum_{y\in B} u_j(z)\,\mu(z)$$

if the inequalities $c_1 < \ldots < c_4, a_1 < \ldots < a_4$

$$a_4 F(x,\delta R) \ge c_4 F(x,R)$$
$$a_3 F(x,\delta R) \le c_3 F(x,R)$$
$$a_2 F(x,\delta R) \le c_2 F(x,R)$$
$$a_1 F(x,\delta R) \ge c_1 F(x,R)$$

are satisfied. In addition we require $c_4 \le \varepsilon$ and $\frac{1}{C_F}(c_3 - c_2)^{\frac{1}{\beta'}} \frac{\delta}{2} \ge \delta^*$. We can see that the following choice satisfies these restrictions. Let $p = C_F(\delta^*)^\beta$ and $q = c_F(\delta^*)^{\beta'}$. Let us choose

$$c_4 = \varepsilon, a_4 = \frac{2q}{p}c_4,$$
$$c_3 < c_4, a_3 = qc_3,$$
$$c_2 < c_3, a_2 = \frac{1}{2}\min\{pc_2, a_3\},$$
$$c_1 = \frac{1}{2}\min\left\{\frac{a_2}{q}, c_2\right\}, a_1 = qc_1.$$

Let us observe that c_1 can be arbitrarily small since $c_4 \le \varepsilon$, and if the sub-solution is not given from an m up to $a_4 F(x,\delta R)$, it can be extended simply by $u_{i+m} = P_i^{B(x,R)} u_m$. ∎

Proof [of Theorem 10.2] Let us fix a set of constants $c_1 < c_2 < c_3 < c_4 = \varepsilon$ as in Lemma 10.1 and apply $PSMV(F)$ for them. Let us apply Lemma 10.1 for δ^* to receive $PMV_{\delta^*}(F)$ on $B = B(x,R)$. As a consequence for $D = B(x,\delta^* R)$, $u_k(y) = h(y)$ we obtain

$$\max_D h \le \frac{C}{\mu(D)} \sum_{y\in D} h(y)\,\mu(y). \tag{10.5}$$

Similarly $PSMV(F)$ yields

$$\min_D h \ge \frac{c}{\mu(D)} \sum_{y\in D} h(y)\,\mu(y). \tag{10.6}$$

The combination of (10.5) and (10.6) gives the elliptic Harnack inequality for the shrinking parameter δ^*. Finally (H) can be shown by using the standard chaining argument along a finite chain of balls. The finiteness of the number of balls follows from volume doubling via the bounded covering principle. ∎

Theorem 10.3. *If (p_0) and (VD) hold, and there is an $F \in W_0$ for which $PMV(F)$ and $PSMV(F)$ are satisfied, then $E \simeq F$ and (TC) is true.*

Proposition 10.1. *Assume (p_0) and (VD). If $PLE(F)$ for $F \in W_0$ holds, then*
$$E(x, R) \geq cF(x, R).$$

Proof   It follows from $PLE(F)$ that there are $c, C, 1 > \delta > \delta' > 0, 1 > \varepsilon > \varepsilon' > 0$ such that for all $x \in \Gamma, R > 1, A = B(x, 2R)$ and $n : \varepsilon' F(x, R) < n < \varepsilon F(x, R), r = \delta' R, y \in B = B(x, r)$

$$\tilde{P}_n^A(x, y) = P_n^A(x, y) + P_{n+1}^A(x, y) \geq \frac{c\mu(y)}{V(x, R)}.$$

It follows for $F = \varepsilon F(x, R)$ and $F' = \varepsilon' F(x, R)$ that

$$E(x, 2R) = \sum_{k=0}^{\infty} \sum_{y \in B(x, 2R)} P_k^A(x, y) \geq \sum_{k=0}^{\infty} \sum_{y \in B} \frac{1}{2} \tilde{P}_k^A(x, y)$$

$$\geq \sum_{k=F'}^{F} \sum_{y \in B} \frac{1}{2} \tilde{P}_k^A(x, y) \geq c\frac{V(x, r)}{V(x, R)} F(x, R) \geq cF(x, R).$$

■

Proposition 10.2. *If $DUE(F)$ holds for an $F \in W_0$, then*

$$\rho(x, 2R)\, v(x, 2R) \leq CF(x, 2R).$$

The first step towards the upper estimate of ρv is to show an upper estimate for λ^{-1}.

Proposition 10.3. *If $(p_0), DUE(F), (VD)$ hold and $F \in W_0$, then*

$$\lambda(x, R) \geq cF^{-1}(x, R). \tag{10.7}$$

Proof   Assume that $C_1 > 1, n = F(x, C_1 R), \ y, z \in B = B(x, R)$. Let us use Lemma 8.8 and $DUE(F)$ to obtain

$$P_{2n}(y, z) \leq C\frac{\mu(z)}{(V(y, f(y, 2n))V(z, f(z, 2n)))^{1/2}}.$$

From (VD) and $F \in W_0$ it follows for $w = y$ or z $d(x, w) \leq R < C_1 R = f(x, n)$ that
$$\frac{V(x, C_1 R)}{V(w, C_1 R)} \leq C,$$

which by using (p_0) yields that for all n,

$$P_n(y, z) \leq C\frac{\mu(z)}{V(x, f(x, n))}.$$

If ϕ is the left eigenvector (measure) belonging to the smallest eigenvalue λ of $I - P^B$ and normalized to $(\phi 1) = 1$, then

$$(1-\lambda)^{2n} = \phi P_{2n}^B 1 = \sum_{y,z \in B(x,R)} \phi(z) P_{2n}^B(z,y) \leq \sum_{y \in B(x,R)} \frac{C\mu(y)}{\min_{z \in B(x,R)} V(z, f(z, 2n))}$$

$$\leq C \max_{z \in B(x,R)} \left(\frac{R}{f(z, 2n)} \right)^\alpha = C \max_{z \in B(x,R)} \left(\frac{1}{C_1} \frac{f(x, 2n)}{f(z, 2n)} \right)^\alpha$$

$$\leq C \left(\frac{1}{C_1} C_f \right)^\alpha \leq \frac{1}{2}$$

if $C_1 = 2C^{1/\alpha} C_f$. Using the inequality and $1 - \xi \geq \frac{1}{2} \log \frac{1}{\xi}$ for $\xi \in [\frac{1}{2}, 1]$, where $\xi = 1 - \lambda(x, R)$, we have

$$\lambda(x, R) \geq \frac{\log 2}{2n} \geq cF(x, C_1 R)^{-1} > cF(x, R)^{-1}.$$

∎

Proof [of Proposition 10.2] Let us recall from (3.9) that

$$\lambda(x, 2R) \rho(x, R, 2R) V(x, R) \leq 1$$

in general and the application of (VD) and (10.7) immediately yields the statement. ∎

Proposition 10.4. *Assume (p_0). If $PLE(F)$ for an $F \in W_0$ holds, then there is a $c > 0$ such that for all $R > 0, x \in \Gamma$*

$$\rho(x, R, 2R) v(x, R, 2R) \geq cF(x, 2R)$$

Proof   The inequality (3.16) establishes that

$$\rho(x, R, 2R) v(x, R, 2R) \geq \min_{z \in \partial B(x, \frac{3}{2}R)} E(z, R/2).$$

From Proposition 10.1 we know that

$$\min_{z \in \partial B(x, \frac{3}{2}R)} E(z, R/2) \geq c \min_{z \in \partial B(x, \frac{3}{2}R)} F(z, R/2),$$

and from $F \in W_0$, it follows that

$$\rho(x, R, 2R) v(x, R, 2R) \geq \min_{z \in \partial B(x, \frac{3}{2}R)} F(z, R/2) \geq cF(x, 2R).$$

∎

Proof [of Theorem 10.3] From Proposition 10.2 we have that $\rho v < CF$ which together with Proposition 10.4, yields that

$$\rho\left(x, R, 2R\right) v\left(x, R, 2R\right) \simeq F\left(x, 2R\right).$$

Since $F \in W_0$, we have that $\rho v \in W_0$. From Proposition 7.1 and from $(aD\rho v)$ the Einstein relation follows:

$$E\left(x, 2R\right) \simeq \rho\left(x, R, 2R\right) v\left(x, R, 2R\right) \simeq F\left(x, 2R\right) \tag{10.8}$$

since (H) is ensured by $PMV(F) + PSMV(F)$. Since $F \in W_0$ and $E \simeq F$, it follows that $E \in W_0$ which includes (TC) and, of course, (wTC) too, and the proof of $8 \Longrightarrow 2$ in Theorem 10.1 is completed. ∎

10.2 Off-diagonal lower estimate

Now we are ready to show the off-diagonal lower estimate $LE(F)$:

$$\widetilde{p}_n\left(x, y\right) \geq \frac{c}{V\left(x, f\left(x, n\right)\right)} \exp\left(-C\left[\frac{F\left(x, d\right)}{n}\right]^{\frac{1}{\beta'-1}}\right),$$

where $d = d\left(x, y\right), F \in W_1$. The proof of the off-diagonal lower estimate uses the modified Aronson's chaining argument. We have shown that

$$(VD) + (TC) + (H) \Longrightarrow DUE\left(E\right), \tag{10.9}$$

$$\left(\overline{E}\right) \Longrightarrow DLE\left(E\right),$$

furthermore for $F \in W_0$,

$$(VD) + DUE\left(F\right) + DLE\left(F\right) + (H) \Longrightarrow NDLE\left(F\right). \tag{10.10}$$

The lower estimate will follow if we show for $F \in W_1$ that

$$(VD) + NDLE\left(F\right) \Longrightarrow LE\left(F\right). \tag{10.11}$$

It results from $(10.9 - 10.11)$ that our final conclusion is

$$(VD) + (TC) + (H) \Longrightarrow LE\left(E\right)$$

if $\beta' > 1$ for E.

Theorem 10.4. *Assume that (Γ, μ) satisfies (p_0). Then for an $F \in W_1$*

$$(VD) + NDLE\left(F\right) \Longrightarrow LE\left(F\right),$$

and if $E \in W_1$

$$(VD) + (TC) + (H) \Longrightarrow LE\left(E\right). \tag{10.12}$$

Let us recall that
$$P_n P_m = P_{n+m}. \tag{10.13}$$
We need a replacement of this property for the operator $\tilde{P}_n$ which is stated below in Lemma 10.5.

Lemma 10.2. *Assume that (p_0) holds on (Γ, μ), then for all integers $n \geq l \geq 1$ such that*
$$n \equiv l \pmod 2, \tag{10.14}$$
we have for all $x, y \in \Gamma$
$$P_l(x, y) \leq C^{n-l} P_n(x, y), \tag{10.15}$$
with a constant $C = C(p_0)$.

Proof   Due to the semigroup property (5.22), we have
$$P_{k+2}(x, y) = \sum_{z \in \Gamma} P_k(x, z) P_2(z, y) \geq P_k(x, y) P_2(y, y).$$

Using (p_0), we obtain
$$P_2(y, y) = \sum_{z \sim y} P(y, z) P(z, y) \geq p_0 \sum_{z \sim y} P(y, z) = p_0,$$

whence $P_{k+2}(x, y) \geq p_0 P_k(x, y)$. Iterating this inequality, we obtain (10.15) with $C = p_0^{-1/2}$. ∎

Lemma 10.3. *Assume that (Γ, μ) satisfies (p_0). Then for all integers $n \geq l \geq 1$ and all $x, y \in \Gamma$,*
$$\tilde{P}_l(x, y) \leq C^{n-l} \tilde{P}_n(x, y), \tag{10.16}$$
where $C = C(p_0)$.

Proof   This is an immediate consequence of Lemma 10.2 because both $P_l(x, y)$ and $P_{l+1}(x, y)$ can be estimated from above via either $P_n(x, y)$ or $P_{n+1}(x, y)$, depending on the parity of n and l. ∎

Remark 10.2. Note that no parity condition is required here in contrast to the condition (10.14) of Lemma 10.2.

Lemma 10.4. *Assume that (Γ, μ) satisfies (p_0). Then for all $n, m \in \mathbb{N}$ and $x, y \in \Gamma$, we have the following inequality*
$$\tilde{P}_n \tilde{P}_m(x, y) \leq C \tilde{P}_{n+m+1}(x, y), \tag{10.17}$$
where $C = C(p_0)$.

Proof   Observe that, by (10.13),
$$\tilde{P}_n \tilde{P}_m = (P_n + P_{n+1})(P_m + P_{m+1}) = P_{n+m} + 2P_{n+m+1} + P_{n+m+2}.$$

By Lemma 10.2, $P_{n+m}(x, y) \leq C P_{n+m+2}$, whence
$$\tilde{P}_n \tilde{P}_m (x, y) \leq C(P_{n+m+1} + P_{n+m+2}) = C \tilde{P}_{n+m+1}.$$

∎

Lemma 10.5. *Assume that (Γ, μ) satisfies (p_0). Then for all $x, y \in \Gamma$ and $k, m, n \in \mathbb{N}$ such that $n \geq km + k - 1$, we have the following inequality*

$$\left(\tilde{P}_m\right)^k (x, y) \leq C^{n-km} \tilde{P}_n(x, y). \tag{10.18}$$

Proof   By induction, (10.17) implies

$$\left(\tilde{P}_m\right)^k (x, y) \leq C^{k-1} \tilde{P}_{km+k-1}(x, y).$$

From inequality (10.16) with $l = km + k - 1$, we obtain

$$\tilde{P}_{km+k-1}(x, y) \leq C^{n-km-(k-1)} \tilde{P}_n(x, y),$$

whence (10.18) follows.   ∎

Proof   [of Theorem 10.4] The proof starts with separation of three cases according to different regions for $d = d(x, y)$.

1. $d(x, y) \leq \delta f(x, n)$,
2. $\delta f(x, n) < d(x, y) \leq \delta' n$,
3. $\delta' n < d(x, y) \leq n$.

In Case 1 $l = n$ by definition and cp_0^{Cn} is a trivial bound. In Case 3

$$\frac{n}{l} \geq QF\left(x, \frac{d}{l}\right) \geq c\left(\frac{n}{l}\right)^2$$

which results in $l > cn$ and again the lower estimate is smaller than $exp(-Cn)$ which can be received from (p_0).

Case 2

The proof uses varying radii for a chain of balls.

Assume that $\delta f(x, n) < d(x, y) < \delta' n$. Consider a shortest path π between x and y, write $d = d(x, y)$,

$$m = \left\lfloor \frac{n}{l(n, R, A)} \right\rfloor - 1, \tag{10.19}$$

$R = f(x, n)$, $S = f(y, n)$, $A = B(x, d+R) \cup B(y, d+S)$. Let $o_1 = x$ and

$$r_1 = \lceil \delta c_0 f(o_1, m) \rceil,$$

and choose $o_2 \in \pi : d(o_1, o_2) = r_1 - 1$ and recursively

$$r_{i+1} = \lceil \delta c_0 f(o_{i+1}, m) \rceil \tag{10.20}$$

and $o_{i+1} \in \pi : d(o_i, o_{i+1}) = r_{i+1} - 1$ and $d(y, o_{i+1}) < d(y, o_i)$. Write $B_i = B(o_i, r_i)$. The iteration ends with the first j for which $y \in B_j$. From $F \in W_0$ and $z_{i+1} \in B_i$ it follows that

$$c_1 \le \frac{f(z_i, m)}{f(z_{i+1}, m)} \le C_2, \tag{10.21}$$

and the from triangle inequality it is evident that

$$d(z_i, z_{i+1}) \le 2r_i + r_{i+1} \le \left(2 + \frac{1}{c_1}\right) \delta c_0 f(z_i, m). \tag{10.22}$$

Here we specify $c_0 = (2 + 1/c_1)^{-1}$. Let us recall the definition of $l = l(n, d, A)$:

$$\frac{n}{l} \ge \max_{z \in A} CE\left(z, \frac{d}{l}\right), \tag{10.23}$$

and taking the inverse, we obtain

$$\min_{z \in A} f\left(z, \frac{1}{C}\frac{n}{l}\right) \ge \frac{d}{l}. \tag{10.24}$$

Let us choose $C > C_F \left(\frac{1}{\delta}\right)^\beta$ in (10.23) (using $F \in W_1$) such that

$$f\left(o_i, \frac{1}{C}\frac{n}{l}\right) \le \delta c_0 f\left(o_i, \frac{n}{l}\right) = r_i.$$

By the definition of j,

$$d > \sum_{i=1}^{j-1} r_i \ge (j-1)\frac{d}{l},$$

consequently, $j - 1 \le l$

$$\left(\tilde{P}_m\right)^j (x, y) \ge \sum_{z_1 \in B_0} \cdots \sum_{z_{j-1} \in B_{j-2}} \tilde{P}_m(x, z_1) \tilde{P}_m(z_1, z_2) ... \tilde{P}_m(z_{j-1}, y).$$

Now we use $NDLE$ to obtain

$$\left(\tilde{P}_m\right)^j (x, y) \ge \sum_{z_1 \in B_0} \cdots \sum_{z_{j-1} \in B_{j-2}} \frac{c\mu(z_1)}{V(x, f(x, m))} \cdots \frac{c\mu(y)}{V(z_{j-1}, f(z_{j-1}, m))}$$

$$\ge c^{j-1} \frac{V(o_1, r_1)}{V(x, f(x, m))} \cdots \frac{V(o_{j-1}, r_{j-1})}{V(z_{j-2}, f(z_{j-2}, m))} \frac{\mu(y)}{V(z_{j-1}, f(z_{j-1}, m))}$$

$$\ge c^{j-1} \frac{\mu(y)}{V(x, f(x, m))} \frac{V(o_1, r_1)}{V(z_2, f(z_2, m))} \cdots \frac{V(o_{j-2}, r_{j-2})}{V(z_{j-1}, f(z_{j-1}, m))}.$$

If we use $(10.20), (10.21)$ and (VD) it follows that

$$\left(\tilde{P}_m\right)^j (x,y) \geq \frac{c^{j-1}\mu(y)}{V(x,f(x,m))} \frac{V(o_1,r_1)}{V\left(z_2,\frac{1}{\delta c_0 c_1}r_1\right)} \cdots \frac{V(o_{j-2},r_{j-2})}{V\left(z_{j-1},\frac{1}{\delta c_0 c_1}r_{j-2}\right)}$$

$$\geq \frac{c^{j-1}\mu(y)}{V(x,f(x,m))}(c')^{j-2}$$

$$\geq \frac{c\mu(y)}{V(x,f(x,n))}\exp[-C(j-1)]$$

$$\geq \frac{c\mu(y)}{V(x,f(x,n))}\exp[-Cl] \tag{10.25}$$

Finally from Lemma 10.5 we know that there is a $c > 0$ such that

$$\tilde{P}_n \geq c^{n-lm}\left(\tilde{P}_m\right)^l$$

if $n \geq lm+l-1$. Let us note that from (10.19) it follows that $n-lm+l \leq 3l$ which results in

$$\tilde{P}_n \geq c^{n-lm}\left(\tilde{P}_m\right)^l \geq c'\frac{c^{3l}}{V(x,f(x,n))}\exp(-Cl)$$

$$\geq \frac{c}{V(x,f(x,n))}\exp\left[-C\left(\frac{F(x,d(x,y))}{n}\right)^{\frac{1}{\beta'-1}}\right].$$

This completes the proof of the lower estimate. ∎

11

Closing remarks

11.1 Parity matters

Let us emphasize that (LE) contains the estimate for $p_n + p_{n+1}$ rather than for p_n. In this section, we discuss to what extent it is possible to estimate $p_n(x, y)$ from below alone (assuming $n \geq d(x, y)$, of course). In general, for a parity reason there is no lower bound for $p_n(x, y)$. Indeed, on any bipartite graph, the length of any path from x to y has the same parity as $d(x, y)$. Therefore, $p_n(x, y) = 0$ if $n \not\equiv d(x, y)$ (mod 2). We immediately obtain the following result for bipartite graphs.

Corollary 11.1. *If (Γ, μ) is bipartite and satisfies $LE(E)$, then*

$$p_n(x, y) \geq \frac{c}{V(x, e(x, n))} \exp\left[-C\left(\frac{E(x, d)}{n}\right)^{\frac{1}{\beta'-1}}\right] \qquad (11.1)$$

for all $x, y \in \Gamma$ and $n \geq 1$ such that

$$n \geq d(x, y) \quad and \quad n \equiv d(x, y) \quad (mod\ 2). \qquad (11.2)$$

Proof Indeed, assuming (11.2), furthermore $n+1$ and $d(x, y)$ have different parity, whence $p_{n+1}(x, y) = 0$, and (11.1) follows from $LE(E)$. ∎

If there is enough "parity mixing" in the graph, then one does get the lower bound regardless of the parity of n and $d(x, y)$.

Corollary 11.2. *Assume that graph (Γ, μ) satisfies $LE(E)$, (p_0) and the following "mixing" condition: there is an **odd** positive integer n_0 such that*

$$\inf_{x \in \Gamma} P_{n_0}(x, x) > 0. \qquad (11.3)$$

Then the lower bound (11.1) holds for all $x, y \in \Gamma$ and $n > n_0$ provided that $n \geq d(x, y)$.

For example if $n_0 = 1$, then the hypothesis (11.3) means that each point $x \in \Gamma$ has a loop edge $x \sim x$. If $n_0 = 3$ and there are no loops, then (11.3) means that for each point $x \in \Gamma$, there is an triangle $x \sim y$, $y \sim z$, $z \sim x$. This property holds, in particular, for the graphical Sierpinski gasket (see Figure 1.1).

Proof   From the semigroup property, for any positive integer m, we obtain

$$p_{2m}(x,x) \geq \frac{1}{V(x,m+1)} \left(\sum_{z \in B(x,m+1)} p_m(x,z)\mu(z) \right)^2 = \frac{1}{V(x,m+1)}.$$

The condition (p_0) and Proposition 2.1 imply $V(x,m+1) \leq C^{m+1}\mu(x)$ whence

$$P_{2m}(x,x) = p_{2m}(x,x)\mu(x) \geq C^{-m-1}.$$

Since we will use this lower estimate only for a bounded range of $m \leq m_0$, we can rewrite it as

$$P_{2m}(x,x) \geq c, \qquad (11.4)$$

where $c = c(m_0) > 0$.

Assuming $n > n_0$, due to the semigroup property (5.22), we have

$$p_n(x,y) = \sum_{z \in \Gamma} p_{n-n_0}(x,z)P_{n_0}(z,y) \geq p_{n-n_0}(x,y)P_{n_0}(y,y) \qquad (11.5)$$

and in the same way:

$$p_n(x,y) \geq p_{n-n_0+1}P_{n_0-1}(y,y). \qquad (11.6)$$

By the hypothesis (11.3), we can estimate $P_{n_0}(y,y)$ from below by a positive constant. Also, $P_{n_0-1}(y,y)$ is bounded below by a constant as in (11.4). Hence, adding up (11.5) and (11.6), we obtain

$$p_n(x,y) \geq c(p_{n-n_0}(x,y) + p_{n-n_0+1}(x,y)). \qquad (11.7)$$

The right-hand side of (11.7) can be estimated from below by $LE(E)$, whence (11.3) follows.   ∎

Finally, let us show an example which explains why we cannot replace in general $p_n + p_{n+1}$ by p_n in $LE(E)$, even though assuming the parity condition $n \equiv d(x,y) \pmod 2$. Let (Γ, μ) be $\mathbb{Z}^D$ with the standard weight $\mu_{xy} = 1$ for $x \sim y$, and let $D > 4$. We modify Γ by adding one more edge ξ of weight 1 which connects the origin $o = (0,0,...,0)$ to the point $(1,1,0,0,...,0)$, and denote the new graph by (Γ', μ').

Clearly, the volume growth and the Green kernel on (Γ', μ') are of the same order as on (Γ, μ), that is

$$V(x,r) \simeq r^D \quad \text{and} \quad g(x,y) \simeq d(x,y)^{2-D}.$$

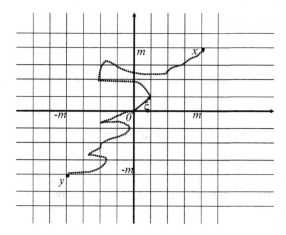

Fig. 11.1. Every path of odd length from x to y goes through o and ξ.

Hence, for both graphs we have

$$p_n(x,y) \le Cn^{-D/2} \exp\left[-\frac{d^2(x,y)}{Cn}\right] \qquad (11.8)$$

and a similar lower bound for $p_n + p_{n+1}$. Since $\mathbb{Z}^D$ is bipartite, for (Γ, μ) we have

$$p_n(x,y) \ge cn^{-D/2} \exp\left[-\frac{d^2(x,y)}{cn}\right]$$

$$\text{if } n \equiv d(x,y) \pmod 2 \quad \text{and} \quad n \ge d(x,y). \qquad (11.9)$$

Let us show that (Γ', μ') does not satisfy (11.9). Fix a (large) odd integer m and consider points $x = (m, m, 0, 0, ..., 0)$ and $y = -x$ (see Figure 11.1).

The distance $d(x,y)$ on Γ is equal to $4m$, whereas the distance $d'(x,y)$ on Γ' is $4m - 1$, due to the shortcut ξ. Write $n = m^2$. Then $n \equiv d'(x,y) \pmod 2$ and $n > d'(x,y)$. Let us estimate from above $p_n(x,y)$ on (Γ', μ'), and show that it does not satisfy the lower bound (11.9). Since n is odd and all odd paths from x to y have to go through the edge ξ, the strong Markov property yields

$$p_n(x,y) = \sum_{k=0}^{n} \mathbb{P}_x(\tau_o = k) p_{n-k}(o,y). \qquad (11.10)$$

If $n - k < m$, then $p_{n-k}(o,y) = 0$. If $n - k \ge m$, then we estimate $p_{n-k}(o,y)$ by (11.8) as follows

$$p_{n-k}(o,y) \le \frac{C}{(n-k)^{D/2}} \le \frac{C}{m^{D/2}}.$$

Therefore, (11.10) implies

$$p_n(x, y) \leq Cm^{-D/2} \mathbb{P}_x \{\tau_o < \infty\}.$$

The $\mathbb{P}_x$-probability of hitting o is of the order $g(x, o) \simeq m^{2-D}$. Hence, we obtain

$$p_n(x, y) \leq Cm^{-(3D/2-2)} = C'n^{(3D/4-1)} = o(n^{-D/2})$$

so that the lower bound (11.9) cannot hold.

Exercise 11.1. Show that if (Γ, μ) satisfies $(p_0), (VD), (TC)$ and $DUE(E)$ for an $F \in W_0$, then there is an $C \geq 1$. such that for all $y \in \Gamma, m \geq n \geq 1, d(x, y) \leq e(y, m)$,

$$c\left(\frac{m}{n}\right)^{\frac{\alpha'}{\beta}} \leq \frac{p_n(x, x)}{p_m(y, y)} < C\left(\frac{m}{n}\right)^{\frac{\alpha}{\beta'}}. \tag{11.11}$$

11.2 Open problems

1. The independence of $(VD), (TC)$ and (H) (or (MV) or its alternatives) is not discussed here. It seems plausible that (TC) and (MV) have a tiny common part. We expect that there is a condition (X) for which $(TC) \Longrightarrow (X)$ holds but the opposite does not, and the triplet of $(VD), (X)$ and (MV) is equivalent to that of $(VD), (TC)$ and (MV).

2. We have seen that the regularity of E is a key getting upper and two-sided heat kernel estimates. It would be desirable to have a set of conditions which are confined to volume, to $\rho(x, R, 2R) v(x, R, 2R)$ and other rough isometry invariant quantities.

3. How can we characterize weighted graphs with a property $(\overline{E}), (wTC)$ or (TC)? Again, volume doubling is a reasonable restriction for this problem, but we had better avoid (H).

Parabolic Harnack inequality

In this chapter a two-sided heat kernel estimate and its equivalence to the parabolic Harnack inequality are shown.

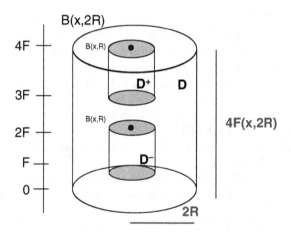

Fig. 12.1. The test balls for PH

Definition 12.1. *We say that* **PH** (F), *the parabolic Harnack inequality holds for a function F if for a weighted graph (Γ, μ) there is a constant $C > 0$ such that for any $x \in \Gamma, R, k > 0$ and any solution $u \geq 0$ of the heat equation*

$$\partial_n u = \Delta u$$

on $\mathcal{D} = [k, k+F(x,R)] \times B(x, 2R)$, the following is true. On smaller cylinders defined by

$$\mathcal{D}^- = [k + \frac{1}{4}F(x, R), k + \frac{1}{2}F(x, R)] \times B(x, R),$$

$$and \; \mathcal{D}^+ = [k + \frac{3}{4}F(x, R), k + F(x, R)) \times B(x, R),$$

and taking $(n_-, x_-) \in \mathcal{D}^-, (n_+, x_+) \in \mathcal{D}^+$,

$$d(x_-, x_+) \leq n_+ - n_-, \tag{12.1}$$

the inequality

$$u_{n_-}(x_-) \leq C\tilde{u}_{n_+}(x_+)$$

holds, where the short notation $\tilde{u}_n = u_n + u_{n+1}$ was used. (See Figure 12.1.)

Let us recall that $F \in V_1$ if there are $\beta' > 1, c_F > 0$ such that for all $R > r > 0, x \in \Gamma, y \in B(x, R)$,

$$\frac{F(x, R)}{F(y, r)} \geq c_F \left(\frac{R}{r}\right)^{\beta'}. \tag{12.2}$$

Theorem 12.1. *If a weighted graph* (Γ, μ) *satisfies* (p_0), *then the following statements are equivalent:*

1. (VD) *holds and there is an* $F \in W_1$ *such that* $g(F)$ *is satisfied;*
2. $(VD), (H)$ *and* (wTC) *hold, furthermore* $E \in V_1$ *or* $\rho v \in V_1$;
3. $(VD), (H)$ *and* (TC) *hold, furthermore* $E \in V_1$ *or* $\rho v \in V_1$;
4. $(VD), (H)$ *and* (ER) *hold, furthermore* $E \in V_1$ *or* $\rho v \in V_1$
5. $(VD), (H)$ *and* $\rho v \in V_1$ *hold;*
6. $(VD), (H)$ *and* $RLE(E)$ *hold, furthermore* $E \in V_1$;
7. (VD) *and* $UE(F), PLE(F), F \in W_1$ *are satisfied;*
8. (VD) *holds and there is an* $F \in W_1$ *such that for any* $0 < \delta \leq 1, PMV(F)$ *and* $PSMV(F)$ *are true;*
9. *there is an* $F \in W_1$ *such that the two-sided heat kernel estimate holds: there are* $C, \beta \geq \beta' > 1, c > 0$ *such that for all* $x, y \in \Gamma, n \geq d(x, y)$

$$c\frac{\exp\left[-C\left(\frac{F(x,d)}{n}\right)^{\frac{1}{\beta'-1}}\right]}{V(x, f(x, n))} \leq \tilde{p}_n(x, y) \leq C\frac{\exp\left[-c\left(\frac{F(x,d)}{n}\right)^{\frac{1}{\beta-1}}\right]}{V(x, f(x, n))} \tag{12.3}$$

where $d = d(x, y)$;
10. *there is an* $F \in W_1$ *such that* $PH(F)$ *holds.*

The equivalence of the statements $1 - 8$ is based on Theorem 10.1. What remains is to incorporate 9 and 10. We will show that the mean value inequalities for $F \in W_1$ are equivalent to the parabolic Harnack inequality and the two-sided heat kernel estimate (12.3). Let us recall Lemma 10.1 which states that under (p_0) and (VD), PMV implies PMV_δ for some test balls. On the basis of this observation we deduce

$$\left.\begin{array}{c} PMV_\delta \\ PSMV \end{array}\right\} \Longrightarrow PH.$$

Remark 12.1. It is known that if the R^2−parabolic Harnack inequality is true for a given profile $\mathcal{C} = \{c_1, c_2, c_3, c_4, \eta\}$, then it is true for all profiles. This follows easily from the bounded covering principle. The bounded covering principle holds on any connected graph (with (p_0)) having the volume doubling property. The only task is to show that the graph can be equipped with a proper metric and measure for which the volume doubling is still true. This is straightforward for R^2 and follows easily for $F(x, R) \in W_1$. Let $f(x, n)$ denote the inverse of $F(x, R)$ and $U = [s_0, s_0 + F(x, R)] \times B(x, R)$. It is clear that the box-metric on the direct product:

$$\rho((s, y), (t, z)) = \frac{1}{2} \left(d(y, z) + |f(x, s - s_0) - f(x, t - s_0)| \right)$$

is suitable for our purposes. Metric balls are $U' = [s_0 + s, s_0 + s + F(y, r)] \times B(y, r)$ of radius r centered at (y, s). The measure is induced by μ, $\nu((y, s)) = \mu(y)$, and

$$\nu(U') = V(y, r) F(y, r)$$

and it has volume doubling property on metric balls thanks to the doubling property of the original volume and $F \in W_1$. Consequently, this setup ensures that the covering principle works on U, and so thus the standard chaining argument can be applied to show that $PH(F)$ holds for all profiles provided that it holds for a given one. There is one delicate point. In order to form a chain between two metric balls with far geometric centers, we need small time gaps which is ensured by $\beta' > 1$.

Exercise 12.1. The details of the chaining argument are left to the reader as an exercise.

Exercise 12.2. Decomposition into Dirichlet solution. Assume that u is a solution on $[0, n] \times B(x, R)$. Let $\varepsilon > 0$,

$$v_n(x) = \sum_{i, y} a_i(y) P_{n-1}^{B(x, R)}(x, y),$$

where the sum runs for $0 < i \leq n, y \in \partial B(x, \varepsilon R)$ or $i = 0, y \in B(x, \varepsilon R)$. Show that as can be defined iteratively such that $a \geq 0, u(n, y) = v(n, y)$ if $y \in B(x, \varepsilon R)$ and $v \leq u$ everywhere.

Lemma 12.1. *Assume* (p_0). *Let* $F \in W_1$, *then*

$$PH(F) \implies DUE(F), DLE(F).$$

Proof   From PH we have that

$$P_k(x, y) \leq C P_{2k}(z, y)$$

if $z \in B(x, r), r = f(x, n)$. Consequently,

$$P_k(x,y) \le \frac{C}{V(x,r)} \sum_{z \in B(x,r)} \mu(z) P_{2k}(z,y)$$

$$= \frac{C}{V(x,r)} \sum_{z \in B(x,r)} P_{2k}(y,z)\,\mu(y)$$

$$\le \frac{C\mu(y)}{V(x,r)}.$$

For the lower bound let us use

$$P_k(z,y) \le C P_{2k}(x,y)$$

for $z \in B(x,s)$, $s = f(x,2k)$. Let u be a solution on $[0,2k] \times B(x,s)$:

$$u_i(z) = \begin{cases} 1 \text{ if } & 0 \le i \le k \\ \sum_{v \in B(x,s)} P_{i-k}(z,v) & \text{if } k < i \le 2k \end{cases}.$$

The PH yields that

$$c = c u_k(x) \le u_{2k}(y) = \sum_{v \in B(x,s)} P_k(y,v)$$

$$= \sum_{v \in B(x,s)} \frac{\mu(v)}{\mu(y)} P_k(v,y) \le C \sum_{v \in B(x,s)} \frac{\mu(v)}{\mu(y)} P_{2k}(x,y)$$

$$= \frac{C V(x,s)}{\mu(y)} P_{2k}(x,y).$$

Let us observe that we have shown a bit more than the diagonal estimate. ∎

Theorem 12.2. *Assume* (p_0). *Let* $F \in W_1$, *then the following equivalence holds:*

$$(VD) + PMV(F) + PSMV(F) \Longleftrightarrow PH(F).$$

Proof   The proof consists of several small steps.

1. Let $\delta^* = \frac{1}{C_F}(c_3 - c_2)^{\frac{1}{\beta'}} \delta/2$ from the parabolic super mean value inequality $PSMV(F)$. By Lemma 10.1, for δ^* and for appropriate c_is, we have $PMV_{\delta^*}(F)$,

$$\max_{\Phi^+} u \le \frac{C}{\nu(\Phi^-)} \sum_{\Phi^-} u_i(z)\mu(z). \tag{12.4}$$

Write $\Phi^+ = [c_3 F, c_4 F] \times B(x, \delta^* R)$ and $\Phi^- = [c_1 F, c_2 F] \times B(x, \delta^* R)$. Let us consider some constants $c_6 > c_5 > c_4$. The parabolic super mean value inequality $PSMV(F)$ with $\mathcal{D}^+ = [c_5 F, c_6 F] \times B(x, \delta^* R), \mathcal{D}^- = \Phi^-$ means that

$$\min_{\mathcal{D}^+} u \ge \frac{c}{\nu(\Phi^-)} \sum_{\Phi^-} u_i(z)\mu(z). \tag{12.5}$$

The combination of (12.4) and (12.5) results in

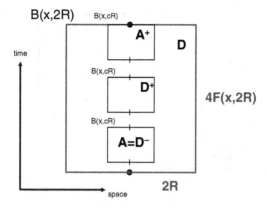

Fig. 12.2. The combination of PMV and $PSMV$

$$\max_{D^-} u \le C \min_{D^+} u, \tag{12.6}$$

which is the parabolic Harnack inequality for Dirichlet solutions with constants $c_3 < c_4 < c_5 < c_6$, δ^*, in other words, $D^- = \Phi^+, D^+ = \mathcal{D}^+$. (See the Figure 12.2.)

2. Let us use the decomposition trick (as in Exercise 12.2) for an arbitrary solution $w \ge 0$ on $D = [0, F(x, R)] \times B(x, R)$. Assume that the maximum on D^+ is attained for k. The nonnegative linear decomposition results in a Dirichlet solution $u \ge 0$ on $\mathcal{D}$ for which $u_k = w_k$ on $B(x, \delta^* R)$ and $u \le w$ in general. Now we use (12.6) and we obtain

$$\max_{D^-} w = \max_{B(x, \delta^* R)} u_k \le C \min_{D^+} u \le C \min_{D^+} w.$$

That means that we have $PH(F)$ for all solutions and for the given c_is and δ^*.

3. If the parabolic Harnack inequality holds for a set of constants, then it is also true for an arbitrary set of constants if $F \in W_1$. The key is that $\beta' > 1$ ensures that the time dimension of the space-time window shrinks faster than its space dimension (See Figure 12.3,12.4, and see Remark 12.1).

4. The implication $PH(F) \implies (VD)$ can be seen along the lines of the classical proof (cf. [31]). The diagonal upper and lower estimates are deduced from PH in Lemma 12.1:

$$p_n(x, x) \le \frac{C}{V(x, f(x, n))}, \tag{12.7}$$

and

$$\widetilde{p}_m(x, x) \ge \frac{c}{V(x, f(x, m))}. \tag{12.8}$$

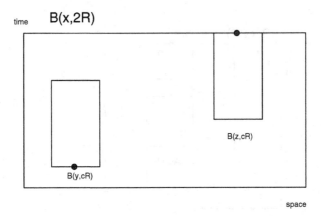

Fig. 12.3. Far points to compare

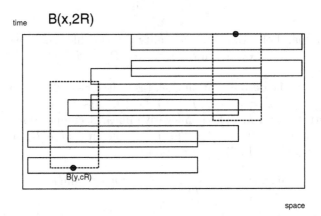

Fig. 12.4. Chain of balls for the comparison

The inequality for $n < cm$:

$$p_n(x, x) \leq C\widetilde{p}_m(x, x) \qquad (12.9)$$

can be obtained from $PH(F)$ with the proper choice of constants. Now let $n = F(x, R), m = F(x, A^p R), p \geq 1$ and $A \geq 2$ be chosen to satisfy $p > \frac{\beta}{\beta'}$ and $A > \left(\frac{C_F}{c_F}\right)^{\frac{1}{p\beta'-\beta}}$. As a result of $(12.7), (12.8)$ and (12.9), we obtain (VD):

$$V(x, 2R) \leq V(x, A^p R) \leq CV(x, R).$$

5. The implication $PH(F) \Longrightarrow PSMV(F)$ is evident. We deduce $PH(F) \Longrightarrow DUE(F)$ as in step 4 and $PMV(F)$ follows from Theorem 10.1.   ■

Remark 12.2. The elliptic Harnack inequality is a direct consequence of the F-parabolic one. This implication follows in the same way as in the classical case.

Proposition 12.1. *From $F \in W_1, UE(F)$ and $LE(F)$ it follows that for a fixed $2 \le K \le \frac{1}{4q} - 1$ there is an $\varepsilon > 0, K\varepsilon < 1$ such that for all $x \in \Gamma, R > 1, A = B(x, R)$,*

$$\widetilde{p}_n^A(w, y) \ge \frac{c}{V(w, K\varepsilon R)}, \tag{12.10}$$

if $F(x, \varepsilon R) \le n \le F(x, K\varepsilon R), w \in B(x, \varepsilon R), y \in B(x, K\varepsilon R), d(w, y) \le n$.

Proof   Let us recall that from $LE(F)$ we have

$$\widetilde{p}_n(w, y) \ge \frac{c}{V(w, K\varepsilon R)} \exp\left[-C\left(\frac{F(w, (1-\varepsilon)R)}{n}\right)^{\frac{1}{\beta'-1}}\right] \tag{12.11}$$

$$\ge \frac{c}{V(w, K\varepsilon R)}.$$

Let us define r by

$$\frac{\mu(w)}{\mu(y)} r_n(w) = \widetilde{P}_n(y, w) - \widetilde{P}_n^A(y, w).$$

It can be decomposed by the exit time and place

$$\frac{\mu(w)}{\mu(y)} r_n(w) \le \sum_{z \in S(x,R)} \sum_{m=0}^{n-1} P_y(T_{x,R} = m, X_m = z)\widetilde{P}_{n-m}(z, w)$$

$$=: \sum_{z \in S(x,R)} \sum_{m=0}^{n-1} \frac{\mu(w)}{\mu(z)} \widetilde{P}_{n-m}(w, z)\pi_m^A(y, z),$$

where it is clear that

$$\sum_{m=0}^{n-1} \sum_z \pi_m^A(y, z) \le 1.$$

From the upper estimate and from $d(w, z) \ge (1 - \varepsilon)R$ with $0 \le m < n$ we have that

$$\frac{1}{\mu(z)} \widetilde{P}_{n-m}(w, z) \le \frac{C}{V(w, f(w, n-m))} \exp\left[-c\left(\frac{F(w, (1-\varepsilon)R)}{n}\right)^{\frac{1}{\beta-1}}\right]$$

$$= \frac{1}{V(w, K\varepsilon R)} \left(\frac{CV(w, K\varepsilon R)}{V(w, f(w, n-m))} \exp\left[-c\left(\frac{F(w, (1-\varepsilon)R)}{n}\right)^{\frac{1}{\beta-1}}\right]\right).$$

We will show that with the right choice of ε, the term in the brackets can be bounded by $c/2$. Let us start with

$$\frac{CV(w, K\varepsilon R)}{V(w, f(w, n-m))} \exp\left[-c\left(\frac{F(w, (1-\varepsilon)R)}{n}\right)^{\frac{1}{\beta-1}}\right]$$

$$\leq C\left(\frac{K\varepsilon R}{f(w, n-m)}\right)^\alpha \exp\left[-c\left(\frac{F(w, (1-\varepsilon)R)}{n}\right)^{\frac{1}{\beta-1}}\right]$$

$$\leq C\exp\left[C\left(\frac{\varepsilon K R}{f(w, n-m)}\right)^\sigma - c\left(\frac{F(w, (1-\varepsilon)R)}{n}\right)^{\frac{1}{\beta-1}}\right]$$

for arbitrary $\gamma > 0, \sigma = \frac{\beta(\beta'-1)}{(\beta-1)^2}$. Now we use Lemma 6.7,

$$k_w(n-m, K\varepsilon R) + 1 \geq c\left(\frac{K\varepsilon R}{f(w, n-m)}\right)^{\frac{\beta}{\beta-1}},$$

furthermore, $k+1 \leq 2k$ if $k \geq 1$ and

$$2k_w(i, r) \leq C\left(\frac{F(w, r)}{i}\right)^{\frac{1}{\beta'-1}}$$

to obtain

$$\left(\frac{r}{f(w, i)}\right)^{\frac{\beta(\beta'-1)}{(\beta-1)^2}} \leq C\left(\frac{F(w, r)}{i}\right)^{\frac{1}{\beta-1}}.$$

We can proceed as follows:

$$C\exp\left[C\left(\frac{K\varepsilon R}{f(w, n-m)}\right)^\sigma - c\left(\frac{(1-\varepsilon)R}{f(w, n-m)}\right)^\sigma\right]$$

$$\leq C\exp\left[\left(\frac{K\varepsilon\alpha C}{(1-\varepsilon)^\sigma} - c\right)\left(\frac{(1-\varepsilon)R}{f(w, n-m)}\right)^\sigma\right]$$

$$\leq C\exp\left[-\frac{c}{2}\left(\frac{(1-\varepsilon)R}{f(w, n)}\right)^\sigma\right].$$

The original restrictions on n and R can be used to receive

$$C\exp\left[\alpha\frac{K\varepsilon R}{f(w, n-m)} - ck\right] \leq C\exp\left(-c\varepsilon^{-\sigma}\right).$$

This can be made arbitrarily small with the right choice of ε, which results in

$$\widetilde{p}_{n-m}(w, z) \leq \frac{c}{2}\frac{1}{V(w, K\varepsilon R)},$$

which, by using (12.11), completes the proof.   ∎

Proposition 12.2. *Assume* (p_0). *Then for an* $F \in W_1$

$$LE\,(F) \Longrightarrow (VD).$$

Proof   Let us fix $n = F(x, R)$, $r = 2R$. $LE\,(F)$ directly provides the following inequality

$$1 \geq \frac{1}{2} \sum_{y \in B(x, 2r)} \widetilde{P}_n(x, y) \tag{12.12}$$

$$\geq \sum_{y \in B(x, 2r)} \frac{c\mu(y)}{V(x, f(x, n))} \exp\left[-C\left(\frac{F(x, 2r)}{n}\right)^{\frac{1}{\beta' - 1}}\right] \tag{12.13}$$

$$= \frac{cV(x, 2r)}{V(x, r)} \exp(-C). $$

∎

Proof   [of Theorem 12.1] First of all, we have seen that under conditions $(VD) + (H) + (*)$

$$E \in W_0 \text{ and } \rho v \in W_0,$$

which, together with (ER) and $E \in V_1$ or $\rho v \in V_1$, implies that that both functions belong to W_1. On the basis of this observation, the equivalence of $1-7$ and 8 is established by Theorem 10.1. The equivalence 8 and 10 is given in Theorem 12.2. In Theorem 8.2 we have seen that the upper estimate $UE\,(F)$ follows from (H) via (MV). It is clear that $PLE \Longrightarrow NDLE$ and in Theorem 10.4 we have seen that for $\beta' > 1$ $(VD) + (TC) + NDLE\,(F) \Longrightarrow LE\,(F)$. For any $F \in W_1$, $LE\,(F)$ implies (VD), as it was shown in Proposition 12.2. Finally, $UE\,(F) + LE\,(F) \Longrightarrow PLE\,(F)$ has been shown in Proposition 12.1. Thus, the equivalence of 7 and 9 is shown and the whole statement is proved.

∎

Remark 12.3. Let us note that we have seen somewhat better upper and lower estimates (which are in fact equivalent to the presented ones). Write $d = d(x, y)$. Following the proof of the upper estimate in Section 8.7, we can see that

$$p_n(x, y) \leq \frac{C \exp\left[-ck_y\left(n, \frac{1}{2}d\right)\right]}{V(x, f(x, n))} + \frac{C \exp\left[-ck_x\left(n, \frac{1}{2}d\right)\right]}{V(y, f(y, n))}. \tag{12.14}$$

The intermediate estimate (10.25) gives a stronger lower bound:

$$\widetilde{p}_n(x, y) \geq \frac{c}{V(x, f(x, n))} \exp\left[-Cl(x, n, A)\right], \tag{12.15}$$

where $A = B(x, e(x, n) + d(x, y)) \cup B(y, e(y, n) + d(x, y)), n \geq d(x, y)$.

12.1 A Poincaré inequality

In this section we show that a Poincaré type inequality follows from the parabolic Harnack inequality.

Remark 12.4. Classical results on two-sided Gaussian estimates are based on the Poincaré inequality (P_2):

$$\sum_{y\in B(x,R)} \mu(y)\left(f(y)-f_B\right)^2 \le CR^2 \sum_{y,z\in B(x,R+1)} \mu_{y,z}(f(y)-f(z))^2, \quad (12.16)$$

where

$$f_B = \frac{1}{V(x,R)} \sum_{y\in B(x,R)} \mu(y)f(y).$$

First Li and Yau [69] showed that the volume doubling property and the Poincaré inequality imply the Gaussian upper bound (GE_2) for certain manifolds. Later Grigor'yan [40] showed the same by eliminating the restrictions on the manifold. Finally Saloff-Coste proved that for manifolds

$$(VD) + (P_2) \Longleftrightarrow (GE_2) \Longleftrightarrow (PH_2).$$

The same statement was proved for weighted graphs by Delmotte [31] by assuming that $p(x,x) \ge \gamma > 0$. Let us mention here that Theorem 12.1 is a generalization of these results in graph settings.

Barlow and Bass show (in [8] and in [9]) that for a space-time scaling function $F(R) = R^\beta \wedge R^{\beta'}, \beta > \beta'$ $PH(F)$ is stabile against rough isometry and with respect of the change of comparable weights $\mu_{x,y} \simeq \mu'_{x,y}$. The proof is based on their result that $PH(F)$ is equivalent to the triplet of (VD), the $F-$Poincaré inequality and a so-called cut-off Sobolev inequality (based on F as well), and these conditions are stable.

Proposition 12.3. *Assume* (p_0). *If* $F \in W_1$, *then* $PH(F)$ *implies the Poincaré inequality* $PI(F)$ *: for any function* f *on* Γ, $x \in \Gamma, R > 0$,

$$\sum_{y\in B(x,R)} \mu(y)\left(f(y)-f_B\right)^2 \le CF(x,R) \sum_{y,z\in B(x,R+1)} \mu_{y,z}(f(y)-f(z))^2,$$

$$(12.17)$$

where

$$f_B = \frac{1}{V(x,R)} \sum_{y\in B(x,R)} \mu(y)f(y).$$

Proof   The proof is an easy adaptation of [30, Theorem 3.11]. We consider the Neumann boundary conditions on $B(x,2R)$, and the new transition probability $P'(y,z) \ge P(y,z)$, and equality holds everywhere inside, and strict inequality holds at the boundary. Consider the operator

$$Qg(y) = \sum P'(y,z)g(z)$$

and $W = Q^K$ where $K = CF(x,R)$. Since $Wg > 0$ is a solution of the parabolic equation on $B(x,2R)$ for $g > 0$, by using the hypothesis we have

$$W(f - Wf(y))^2(y) \geq \sum_{z \in B(x,R+1)} \frac{c\mu(z)}{V(y,2R)}(f(z) - Wf(y))^2$$

$$\geq \frac{c}{V(x,3R)} \sum_{z \in B(x,R)} \mu(y)(f(y) - f_{B(x,R)})^2$$

since $\sum_{z \in B(x,R)} \mu(y)(f(y) - \lambda)^2$ is minimal at $\lambda = f_B$. We arrive at

$$\sum_{y \in B(x,R)} \mu(y)(f(y) - f_{B(x,R)})^2$$

$$\leq c \sum_{y \in B(x,R+1)} W(f - Wf(y))^2(y)$$

$$\leq c\left(\|f\|_2^2 - \|Wf\|_2^2\right)$$

$$\leq cK \|\nabla f\|_2^2,$$

where $\|f\|_2^2 = \sum_{y \in B(x,2R)} \mu'(y)f(y)^2$ and $\|\nabla f\|_2^2 = \sum_{y,z \in B(x,2R)} \mu_{y,z}(f(y) - f(z))^2$. The last inequality is the result of the repeated application of

$$\|Wf\|_2^2 \leq \|f\|_2^2 \text{ and } \|f\|_2^2 - \|Wf\|_2^2 \leq \|\nabla f\|_2^2$$

By recalling the definition of K, the result follows . ∎

Proposition 12.4. *Assume* $(p_0), (VD)$. *If* $F \in W_0$, *then* $PI(F)$, *the Poincaré inequality implies a resistance upper bound. There is a* $C > 0$ *such that for any function* $x \in \Gamma, R > 0$,

$$\rho(x,R,2R) \leq C\frac{F(x,2R)}{V(x,2R)}.$$

Proof   Let h be the capacity potential between $A = B(x,R)$ and $D = \partial B(x,2R)$, $h|_A = 1, h|_D = 0$ and $h|_{B(x,3R)\backslash B(x,2R)} = 0$. Let $U = B(x,3R), y \in B = B(x,2R) : d(x,y) = 5/2R$. We know that $V(y,R/2) > cV(x,R)$. Let us observe that $|h - h_U| > 1/2$ either on A or on $B(y,R/2)$. The Poincaré inequality yields that

$$V(x,R) \leq C\sum_{z \in U}(h(z) - h_U)^2 \leq CF(x,R)\sum_{y,z \in U} \mu_{y,z}(h(y) - h(z))^2$$

$$= \frac{CF(x,R)}{\rho(x,R,2R)}.$$

∎

13
Semi-local theory

The (sub-)Gaussian estimate in the classical (fractal, and polynomial growth) case does not depend on the particular location x or y, but on their distance between them. In particular, in the case of $E(x, R) \simeq R^\beta$, for many fractal type graphs, we have

$$\widetilde{p}_n(x, y) \simeq \frac{\exp\left[-c.\left(\frac{d(x,y)^\beta}{n}\right)^{\frac{1}{\beta-1}}\right]}{V\left(x, n^{1/\beta}\right)},$$

where $c.$ is different for the upper and lower estimate. This observation supports the hope that we can obtain two-sided sub-Gaussian estimates in which the exponents are the same up to the constant if the mean exit time is uniform in the space. This is the case in the classical results, where $E(x, R) \simeq R^2$ is the space-time scaling function. The scaling function $F(R) \simeq E(x, R)$ inherits several properties from the mean exit time, and below we collect them together with the properties of the kernel function $m(n, d(x, y))$.

13.1 Kernel function

Let us recall that the kernel function m is defined as follows: $R \wedge n \geq m = m(n, R)) \geq 1$ is the maximal integer for which

$$\frac{n}{m} \leq qF(\frac{R}{m}), \tag{13.1}$$

or $m = 1$ by definition if $n > F(R)$, or there is no appropriate m. Here q is a small fixed constant.

Definition 13.1. *We define a set of uniform scaling functions U_1. $F \in U_1$ if $F : \mathbb{R} \to \mathbb{R}$ and*
1. there are $\beta > 1, \beta' \geq 1, c_F, C_F > 0$ such that for all $R > r > 0, x \in \Gamma, y \in B(x, R)$,

$$c_F \left(\frac{R}{r} \right)^{\beta'} \leq \frac{F(R)}{F(r)} \leq C_F \left(\frac{R}{r} \right)^{\beta}, \tag{13.2}$$

2. there is a $c > 0$ such that for all $x \in \Gamma, R > 0$,

$$F(R) \geq cR^2, \tag{13.3}$$

3.

$$F(R+1) \geq F(R) + 1 \tag{13.4}$$

for all $R \in \mathbb{N}$.

The following lemma provides estimates of the sub-Gaussian kernel function.

Lemma 13.1. *If $F \in U_1$, then for $m = m(n, R)$,*

$$m + 1 \geq c \left(\frac{F(R)}{n} \right)^{\frac{1}{\beta-1}}, \quad m + 1 \geq c' \left(\frac{R}{f(n)} \right)^{\frac{\beta}{\beta-1}}, \tag{13.5}$$

and

$$m \leq C \left(\frac{F(R)}{n} \right)^{\frac{1}{\beta'-1}}, \quad m \leq C \left(\frac{R}{f(n)} \right)^{\frac{\beta'}{\beta'-1}}. \tag{13.6}$$

Proof   The statements easily follow from $F \in U_1$ and from the definition of m. ∎

Lemma 13.2. *In general,*

1. $m(n, R)$ is non-increasing in n and non-decreasing in R,
2. for any $L \in \mathbb{N}$

$$Lm(n, R) \leq m(Ln, LR)$$

Proof   The statements are direct consequences of the definition of m. ∎

13.2 Two-sided estimate

In this section, we present the two-sided sub-Gaussian estimate:

$$p_n(x, y) \leq \frac{C}{V(x, f(n))} \exp\left[-cm(n, d(x, y))\right], \tag{13.7}$$

$$\widetilde{p}_n(x,y) \geq \frac{c}{V(x,f(n))} \exp\left[-Cm(n,d(x,y))\right]. \tag{13.8}$$

Let us recall our estimates in a concise form. We have seen (cf. Remark 12.3) that under $(VD), (TC)$ and (MV) (the latter one follows from (H)),

$$p_n(x,y) \leq \frac{C \exp\left[-ck_y\left(y,n,\frac{1}{2}d\right)\right]}{V(x,e(x,n))} + \frac{C \exp\left[-ck_x\left(n,\frac{1}{2}d\right)\right]}{V(y,e(y,n))}. \tag{13.9}$$

In Theorem 10.4 it was shown that the lower estimate

$$\widetilde{p}_n(x,y) \geq \frac{c}{V(x,e(x,n))} \exp\left[-Cl(x,n,A)\right] \tag{13.10}$$

follows from $(VD), (TC)$ and (H).

The following result is a direct consequence of Theorem 12.1 in the "time homogeneous" case.

Theorem 13.1. *If a weighted graph (Γ,μ) satisfies (p_0), then the following statements are equivalent:*

1. *$(VD), (E)$ and $g(E)$ hold;*
2. *$(VD), (H), (E)$ and (wTC) hold;*
3. *$(VD), (H), (E)$ and (TC) hold;*
4. *$(VD), (H), (E)$ and (ER) hold;*
5. *$(VD), (H), (\rho v)$ and $(aD\rho v)$ hold;*
6. *$(VD), (H), (\rho v)$ and $RLE(E)$ hold;*
7. *(VD) and $UE(F), PLE(F), F \in U_1$ are satisfied;*
8. *(VD) holds, and there is an $F \in U_1$ such that $PMV(F)$ and $PSMV(F)$ are true;*
9. *there is an $F \in U_1$ such that the two-sided heat kernel estimate holds, and there are $C, c > 0$ such that for all $x, y \in \Gamma$, $n \geq d(x,y)$;*

$$c\frac{\exp\left[-Cm(n,d)\right]}{V(x,f(n))} \leq \widetilde{p}_n(x,y) \leq C\frac{\exp\left[-cm(n,d)\right]}{V(x,f(n))},$$

   where $d = d(x,y)$;
10. *there is an $F \in U_1$ such that $PH(F)$ holds.*

Remark 13.1. Let us mention here that under $(p_0), (VD)$ and (H) the uniformity of the mean exit time in the space

$$E(x,R) \simeq F(R) \tag{13.11}$$

ensures that E satisfies the right hand side of (7.20) with a $\beta' > 1$. This explains that in the "classical" case when (13.11) holds, it should not be assumed that $\beta' > 1$, since it follows from the conditions (see Proposition 7.6).

Remark 13.2. It is not immediate but elementary to deduce from (12.14) and (12.15) a particular case of Theorem 12.1 if

$$E\left(x,R\right) \simeq F\left(R\right),$$

or

$$\rho\left(x,R,2R\right)v\left(x,R,2R\right) \simeq F\left(R\right).$$

The key observation is that under $(p_0), (VD), (H)$ and (E),

$$E \in W_0 \Longrightarrow E \in W_1,$$

which means that under the corresponding conditions (E) implies $\beta' > 1$ (c.f. Proposition 7.6). The statements $1-8$ and 10 of Theorem 12.1 are immediate, the two-sided heat kernel estimate

$$c\frac{\exp\left[-Cm\left(n,d\left(x,y\right)\right)\right]}{V\left(x,f\left(n\right)\right)} \leq \widetilde{p}_n\left(x,y\right) \leq C\frac{\exp\left[-cm\left(n,d\left(x,y\right)\right)\right]}{V\left(x,f\left(n\right)\right)} \qquad (13.12)$$

for $F \in W_1$ needs some preparation. It follows from (12.14) and (12.15) and from the fact that for any fixed $C_i > 0, x \in \Gamma$,

$$k\left(x,C_1n,C_2R\right) \simeq l\left(x,C_3n,C_4R\right) \simeq m\left(C_5n,C_6R\right).$$

Remark 13.3. In the particular case when $E\left(x,R\right) \simeq R^\beta$, we recover the sub-Gaussian estimate:

$$c\frac{\exp\left[-C\left(\frac{d^\beta\left(x,y\right)}{n}\right)^{\frac{1}{\beta-1}}\right]}{V\left(x,n^{\frac{1}{\beta}}\right)} \leq \widetilde{p}_n\left(x,y\right) \leq C\frac{\exp\left[-c\left(\frac{d^\beta\left(x,y\right)}{n}\right)^{\frac{1}{\beta-1}}\right]}{V\left(x,n^{\frac{1}{\beta}}\right)}, \qquad (13.13)$$

which is usual for the simplest fractal-like graphs.

13.3 Open problems

This problem has some historical background. It is related to the classical potential theory. Let us recall the notion of normal (Markov) chains which form a sub-class of recurrent chains (cf. [64]). Very briefly, in our setting a reversible Markov chain is normal if for

$$G^n\left(y,x\right) = \sum_{i=0}^n P_i\left(y,x\right),$$

the function

$$K^n\left(y,x\right) = G^n\left(x,x\right) - G^n\left(y,x\right)$$

has a limit as n tends to infinity. Let us recall the definition of the local Green function and the resolvent:

$$G^R(y,x) = G^{B(x,R)}(y,x),$$

and for $\lambda > 0$,

$$G_\lambda(y,x) = \sum_{i=0}^{\infty} e^{-\lambda i} P_i(y,x).$$

Let us define the following functions:

$$K^R(y,x) = G^R(x,x) - G^R(y,x),$$

and

$$K_\lambda(y,x) = G_\lambda(x,x) - G_\lambda(y,x).$$

What is the connection between the convergences of the K functions as n, R or λ^{-1} tends to infinity? Is it true that all of them converge if any of them does? Is it true that K^R converges if and only if the elliptic Harnack inequality holds on Γ? This is the case for normal random walks on $\mathbb{Z}^d$ studied in [88].

List of lettered conditions

(aVD) anti-doubling for volume, page 10

(BC) bounded covering principle, page 12

(DG) Davies-Gaffney inequality, page 127

$DLE(E)$ diagonal lower estimate, page 73

$DLE(F)$ diagonal lower estimate with respect to F, page 74

$(DUE_{\alpha,\beta})$ diagonal upper estimate, page 61

(DUE_ν) diagonal upper estimate with polynomial decay, page 62

(E_β) polynomial mean exit time, page 3

$(\overline{E})$ condition E-bar, page 13

(ER) Einstein relation, page 83

$(FK\rho)$ isoperimetric inequality for resistance, page 116

(FK) Faber-Krahn inequality, page 116

(FKE) isoperimetric inequality for E, page 116

(FK_ν) Faber-Krahn inequality, page 62

$(g_{0,1})$ Green kernel upper bound, page 97

$(GE_{\alpha,\beta})$ two-sided sub-Gaussian estimate, page 4

$g(F)$ two-sided bound on Green kernel, page 89

(H) elliptic Harnack inequality, page 35

$HG(U, M)$ annulus Harnack inequality for Green functions, page 36

$HG(M)$ annulus Harnack inequality for Green functions on balls, page 36

(wHG) weak Harnack inequality for Green functions, page 36

$LE(F)$ lower estimate, page 159

(MV) mean-value inequality, page 96

(MVG) mean-value inequality for G, page 97

$NDLE(F)$ near diagonal lower estimate, page 136

(p_0) controlled weights condition, page 8

$PH(F)$ parabolic Harnack inequality, page 169

$PI(F)$ Poincaré inequality, page 178

$PLE(E)$ particular lower estimate, page 131

$PMV(F)$ parabolic mean-value inequality with $\delta = 1$, page 96

$PMV_\delta(F)$ parabolic mean-value inequality with $\delta < 1$, page 96

$PSMV(F)$ parabolic super mean-value inequality, page 131
$wPMV(F)$ weak parabolic mean-value inequality, page 96
$wPSMV(F)$ weak parabolic super mean-value inequality, page 132
$PUE(E)$ particular upper estimate, page 99
$RLE(F)$ resistance lower estimate, page 87
(ρv) uniform scaling function, page 17
$(aD\rho v)$ anti-doubling for ρv, page 86
VSR very strong recurrence, page 147
(TC) time comparison principle, page 14
(TD) time doubling property, page 14
(wTC) weak time comparison principle, page 14
$UE(E)$ upper estimate, page 99
(V_α) polynomial volume growth, page 10
(VC) volume comparison principle, page 10
(VD) volume doubling property, page 10
(wVC) weak volume comparison principle, page 10
$(*)$ set of conditions equivalent to (ER), page 154

Subject index

References

1. Alexander, S. Orbach, R.: Density of states on fractals, "fractions",. J. Physique (Paris) Lett. **43**, L625-L631, (1982)
2. Aronson, D.G.: Non-negative solutions of linear parabolic equations. Ann. Scuola Norm. Sup. Pisa cl. Sci (3) **22**, (1968), 607-694; Addendum **25**, (1971), 221-228.Barlow, M.T. Random Walks and Diffusion on Fractals, Proc. Int. Congress Math. Kyoto, (1990)
3. Barlow, M.T.: Which values of the volume growth and escape time exponent are possible for a graph? Revista Math. Iberoamericana **20**, 1-31, (2004)
4. Barlow, M.T.: Some remarks on the elliptic Harnack inequality, preprint
5. Barlow, M.T., Bass, F.R.: The Construction of the Brownian Motion on the Sierpinski Carpet, Ann. Inst. H. Poincaré, **25**, 225-257, (1989)
6. Barlow, M.T., Bass, F.R.: Brownian motion and harmonic analysis on Sierpinski carpets, Canadian J. Math., **51**, 673-744, (1999)
7. Barlow, M.T., Bass, F.R.: Divergence form operators on fractal-like domains, J. Func. Anal. **175**, 214-247, (2000)
8. Barlow, M.T., Bass, F.R.: Stability of the Parabolic Harnack Inequality, T. Am. Math. Soc **356,** 4, 1501-1533, (2004)
9. Barlow, M.T., Bass, R., Kumagai, T.: Stability of parabolic Harnack inequality, preprint
10. Barlow, M.T., Coulhon, T., Grigor'yan A.: Manifolds and graphs with slow heat kernel decay, Invent. Math. **144**, 609-649, (2001)
11. Barlow M.T., Hambly, B.: Transition density estimates for Brownian motion on scale irregular Sierpinski gaskets, Ann. IHP **33**, 531-557, (1997)
12. Barlow, M.T.; Nualart, D.: Diffusion on Fractals. in: Lectures on probability theory and statistics. Lectures from the 25th Saint-Flour Summer School held July 10–26, 1995. Edited by P. Bernard. Lecture Notes in Mathematics, 1690. Springer-Verlag, Berlin, (1998)
13. Barlow, M.T., Perkins, E.A.: Brownian Motion on the Sierpinski Gasket, Probab. Th. Rel. Fields, **79**, 543-623, (1988)
14. Barlow, M.T., Taylor, S.J.: Defining fractal Subsets of Z^d, Proc. London Math. Soc. **64**, 125-152. [3] (1991)
15. Benjamini, I., Peres, Y.: Tree-indexed random walks on groups and first passage percolation. (With I. Benjamini). Probab. Theory Rel. Fields. **98**, 91-112, (1994)

16. Bollobás, B.: Random Graphs, Academic Press, London, (1985)
17. Boukricha A.: Das Picard-Prinzip und verwandte Fragen bei Störung von harmonischen Räumen, Math. Ann. **239**, 247-270, (1979)
18. Carne, T.K.: A transmutation formula for Markov chains, Bull. Sci. Math.,(2), **109**, 399-405, (1985)
19. Carron, G.: Inégalités isopérimétriques de Faber-Krahn et conséequences, Actes de la table ronde de géométrie différentielle (Luminy, 1992), Collection SMF Séminaires et Congrées} **1**, 205–232, (1996)
20. Chavel, I.: Isoperimetric Inequalities : Differential Geometric and Analytic Perspectives (Cambridge Tracts in Mathematics, No 145), (2001)
21. Ashok K. Chandra, Prabhakar Raghavan, Walter L. Ruzzo, Roman Smolensky, Prasoon Tiwari: ACM Symposium on Theory of Computing, (1989)
22. Cheeger, J.: Differentiability of Lipschitz functions on Metric Measure Spaces. Geom. Funct. Anal. **9**, 3, 428-517, (1999)
23. Cheeger J., Yau, S.-T.: A lower bound for the heat kernel. Comm. Pure Appl. Math., **34**, 4,465-48, (1981)
24. Chung, F.R.K.: Spectral Graph Theory CBMS Regional Conference Series in Mathematics, 92. Published for the Conference.
25. Coulhon, T.: Analysis on infinite graphs with regular volume growth, JE 2070, No 17/18, November 1997, Université de Cergy-Pontoise
26. Coulhon, T.: Ultracontractivity and Nash type inequalities, J. Funct. Anal., **141:1**, 81-113, (1996)
27. Coulhon, T., Grigor'yan, A.: Random walks on graphs with regular volume growth, Geometry and Functional Analysis, **8**, 656-701, (1998)
28. Grigor'yan, A. Coulhon, T.: Pointwise estimates for transition probabilities of random walks on infinite graphs, In: Trends in Math., Fractals in Graz 2001 (P. Grabner and W. Woess (eds.)), Birkhauser, (2002)
29. Coulhon, T., Saloff-Coste, L.: Variétés riemanniennes isométriques à l'infini, Revista Matemática Iberoamericana, **11**, 3, 687-726, (1995)
30. Davies, E.B.: Heat kernels and spectral theory, Cambridge University Press, Cambridge, (1989)
31. Delmotte, T.: Parabolic Harnack inequality and estimates of Markov chains on graphs. Revista Matemática Iberoamericana **1**, 181–232, (1999)
32. Doyle, P.G.; Snell, J.L.: Random walks and electric networks Carus Mathematical Monographs, 22. Mathematical Association of America, Washington, DC, 1984.
33. Einstein, A.: Ann. Phys. **11**. 170, and **17**, 549,1903.
34. Fabes, E., Stroock, D.: A new proof of the Moser's parabolic Harnack inequality using the old ideas of Nash, Arch. Rat. Mech. Anal., **96**, 327-338, (1986)
35. Fukushima, M.: Dirichlet forms and Markov processes. North Holland Kodansh, 1980.
36. Fukushima, M.; Oshima, Y.; Takeda M.: Dirichlet forms and symmetric Markov Processes de Gruyter Studies in Mathematics, 19. Walter de Gruyter & Co., Berlin, (1994)
37. Goldstein, S.: Random walk and diffusion on fractals, Lect. Notes IMA, 8, Ed. H. Kesten, (1987)
38. Grigor'yan, A.: The heat equation on non-compact Riemannian manifolds, (in Russian) Matem. Sbornik **182:1**, 55-87 Engl. transl., Math. USSR db. **72:1**, 47-77, (1992)

39. Grigor'yan, A.: Gaussian upper bounds for the heat kernel on arbitrary manifolds. J. Differential Geometry **45**, 33-52, (1997)

40. Grigor'yan, A.: Heat kernel upper bounds on a complete non-compact manifold, Revista Math. Iberoamericana **10**, 2, 395-452, (1994)

41. Grigor'yan, A.: Isoperimetric inequalities and capacities on Riemannian manifolds, Operator Theory: Advances and Applications, **109**, 139-153 (Special volume dedicated to V. G. Maz'ya), (1999)

42. Grigor'yan, A.: Estimates of heat kernels on Riemannian manifolds, in "Spectral Theory and Geometry. ICMS Instructional Conference, Edinburgh, 1998", ed. B. Davies and Yu. Safarov, Cambridge Univ. Press, London Math. Soc. Lecture Notes 273 140-225, (1999)

43. Grigor'yan, A.: Gaussian upper bounds for the heat kernel on arbitrary manifolds. J. Differential Geometry **45**, 33-52, (1997)

44. Grigor'yan, A.: Analytic and geometric background of recurrence and non-explosion of the Brownian motion on the Riemannian manifolds. Bull. Amer. Math. Soc. (N.S.) **36**, 2, 135–249, (1999)

45. Grigor'yan, A.: Heat kernel upper bounds on fractal spaces, preprint

46. Grigor'yan, A., Saloff-Coste, L. Dirichlet heat kernel in the exterior of compact set, preprint

47. Grigor'yan, A., Saloff-Coste, L.: Some new examples and stability results concerning Harnack inequalities, preprint

48. Grigor'yan, A., Telcs, A.: Sub-Gaussian estimates of heat kernels on infinite graphs, Duke Math. J., **109**, 3, 452-510, (2001)

49. Grigor'yan, A., Telcs, A.: Harnack inequalities and sub-Gaussian estimates for random walks, to appear in Math. Annal. **324**, 521-556, (2002)

50. Grigor'yan, A., Telcs, A.: Heat kernel estimates on measure metric spaces (in preparation)

51. Gromov M.: Groups of polynomial growth and expanding maps. Publ. Math. Inst. H. Poincaré Probab. Statist. **53**, 57-73, (1981)

52. Harris, T.: The Theory of Branching Processes, Springer, (1963)

53. Hambly, B.M.: Brownian motion on a random recursive Sierpinski gasket. Ann. Probab. **25**, 3, 1059–1102, (1997)

54. Hambly,B., Kumagai, T.: Heat kernel estimates for symmetric random walks on a class of fractal graphs and stability under rough isometries, roc. of Symposia in Pure Math. 72, Part 2, pp. 233–260, Amer. Math. Soc. (2004)

55. Holopainen, I.: Volume growth, Green's functions and parabolicity of ends, Duke Math. J. **97**, 2, 319-346, (1999)

56. Hebisch W.; Saloff-Coste, L.:, Gaussian estimates for Markov chains and random walks on groups, Ann. Probab., **21**, 673-709, (1993)

57. Hebisch W., Saloff-Coste, L.: On the relation between elliptic and parabolic Harnack inequalities, Ann. Inst. Fourier **51**, 5, 1437-1481, (2001)

58. Hughes, B.D.: Random Walks and Random Environments, Vol. 1, Random Walks, Claredon Press (1995)

59. Hughes, B.D.: Random Walks and Random Environments, Vol. 2, Random Environments, Claredon Press (1996)

60. Jones, O.D.: Transition probabilities for the simple random walk on the Sierpinski graph, Stoch. Proc. Appl., **61**, 42-69, (1996)

61. Kesten, H.: Symmetric Random Walks on groups, Trans. Amer. Math. Soc., **92**, 336-354, (1959)

62. Kesten, H.: Sub-diffusive behavior of random walks on a random cluster, Ann. Inst. H. Poincaré **22**, 425-487, (1986)

63. Kesten, H., Spitzer, F.: Random Walks on countably infinite Abelian groups, Acta Math., **114**, 257-267, (1965)

64. Kemény, J.G., Snell, J.L., Knapp, A.: Denumerable Markov Chains, Springer NY., 2. ed., (1976)

65. Kigami, J.: Analysis on fractals, Cambridge Univ. Press, 226 s. - (Cambridge tracts in mathematics ; 143) (2001 - viii)

66. Kusuoka, S. A diffusion process on a fractal, Symposium on Probabilistic Methods in Mathematical Physics, Taniguchi, Katata, 251-274, Academic Press, Amsterdam, (1987)

67. Kusuoka, S., Zhou, X.Y.: Dirichlet forms on Fractals; Poincaré constant and resistance, Probab. Theo. and Rel. F., **93,** 169-196, (1992)

68. Ledoux, M.: The geometry of Markov diffusion generators, lecture notes preprint, (1998)

69. Li, P., Yau, S.-T.: On the parabolic kernel of the Schrödinger operator , Acta Math. **156**, 153-201, (1986)

70. Li, P., Wang, J.: Mean value inequalities, Indiana Univ. Math., J., **48**, 4, 1257-1283, (1999)

71. Moser, J.: On Harnack's Theorem for elliptic differential equations, Communications of Pure and Applied Mathematics, **16**, 101-134, (1964)

72. Moser, J.: On Harnack's theorem for parabolic differential equations, Communications of Pure and Applied Mathematics, **24**, 727-740, (1971)

73. Mandelbrot, B.B.: Fractals: Form, Chance and Dimension, Freemann, San Francisco, (1997)

74. Mathieu, P.: Hitting times and spectral gap inequalities. Ann. Inst. H. Poincaré Probab. Statist. **33**, 4, 437–465, (1997)

75. Mathieu, P.: Inégalités de Sobolev et Temps dáAtteinte, Potential Analysis, 9, 293-300, (1998)

76. McDonald, P.: Isoperimetric conditions, Poisson problems and diffusions in Riemannian manifolds, Potential Analysis **16**, 115-138, (2002)

77. Mosco, U.: Dirichlet form and self-similarity, A.M.S. /IP Studies in Advanced Mathematics 8 (1998)

78. Nash-Williams, C.St.J.A.: Random Walks and electric currents in networks, Proc. Cambridge Phil. Soc., **55**, 181-194, (1958)

79. Osserman, R.: The isoperimetric Inequality, Bull. A.M.S. **84:6**, 1181-1238, (1978)

80. Pólya, G., Szegő, G.: Isoperimetric inequalities in mathematical physics, Princeton University Press, (1951)

81. Pólya, G.: Über eine Aufgabe der Wahrscheinlichkeitstheorie betreffend die Ihrfart in Strassennetz, Math. Annales, 149-160, (1921)

82. Révész, P.: Random Walk in Random & Non-Random Environments, World Scientific Publishing Company, Incorporated, (1990)

83. Röckner, M.: Dirichlet forms on infinite-dimensional "manifold-like" state spaces: a survey of recent results and some prospects for the future. Probability towards 2000 (New York, 1995), 287–306, Lecture Notes in Statist., **128**, Springer, New York, (1998)

84. Rammal, R.: Toulouse, G. Random Walks on fractal structures and percolation clusters, J. Phys. Lett., Paris, **44**, L13-L22, (1983)

85. Saloff-Coste, L.: Isoperimetric Inequalities and decay of iterated kernels for almost-transitive Markov chains Combinatorics Probability and Computing **4**, 419-442, (1995)

86. Saloff-Coste, L.: A note on Poincaré, Sobolev and Harnack inequalities. Duke Math. J. IMRN **2**, 27-38, (1992)

87. Saloff-Coste, L.: Isoperimetric inequalities and decay of iterated kernels for almost-transitive Markov chains, Comb., Probab. & Comp. **4**, 419 - 442, (1995)

88. Spitzer, F.: Principles of Random Walks, Van Nostrand, (1964)

89. Sturm, K-Th.: Diffusion processes and heat kernels on metric spaces, Ann. Probab., **26:1**, 1-55, (1998)

90. Stannat, W.: The theory of generalized Dirichlet forms and its application in analysis and stochastic Mem. Amer. Math. Soc. **142**, 678, (1999)

91. Sturm, K-Th.: Diffusion processes and heat kernels on metric spaces. Ann. Probab. **26:1**, 1-55, (1998)

92. Sung, C-J.: Weak super mean value inequality Proc. Amer. Math. Soc. **130**, 3401-3408, (2002)

93. Stroock, D.W., Varadhan, S.R.S.: Multidimensional diffusion processes. Springer Berlin, (1979)

94. Telcs, A.: Random Walks on Graphs, Electric Networks and Fractals, Probab. Theo. and Rel. Fields, **82**, 435-449, (1989)

95. Telcs, A.: Fractal Dimension and Martin Boundary of Graphs, Studia Sci. Math. Hung. **37**, 143-167, (2001)

96. Telcs, A.: Local Sub-Gaussian Estimates on Graphs: The Strongly Recurrent Case, Electronic Journal of Probability, **6**, 1-33, (2001)

97. Telcs, A.: A note on Rough isometry invariance of resistances, to appear in Comb. Probab. Comput., **11**, 1-6, (2002)

98. Telcs, A.: Volume and time doubling of graphs and random walk, the strongly recurrent case, Communication on Pure and Applied Mathematics, **LIV**, 975-1018, (2001)

99. Telcs, A.: Random walks on graphs with volume and time doubling, Revista Mat. Iber. **22**, 1 (2006)

100. Telcs, A.: Some notes on the Einstein relation, to appear in J. Stat. Phys.

101. Varopoulos, N. Th. Isoperimetric Inequalities for Markov chains, J. Functional Analysis **63**, 215-239, (1985)

102. Varopoulos, N. Th.: Hardy-Littlewood theory for semigroups, J. Funct. Anal., **63**, 215-239, (1985)

103. Varopoulos,N. Th., Saloff-Coste, L., Coulhon, Th.: Analysis and geometry on Groups, Cambridge University Press, (1993)

104. Virág, B.: On the speed of random walks on graphs, Ann. Probab. **28**, 1, 379–394, (2000)

105. Wang, F-Y.: Functional inequalities, semigroup properties and spectrum estimates. Infinite Dimensional Analysis, Quantum Probability and Related Topics, **3**, 2, 263-295, (2000)

106. Weaver, N.: Lipschitz Algebras, World Scientific Press, (1999)

107. Woess, W.: Random walks on infinite graphs and groups, Cambridge University Press, Cambridge, (2000)

108. Zhou, Z.Y.: Resistance dimension, Random Walk dimension and Fractal dimension, J. Theo. Probab. **6**,4,635-652, (1993)

Lecture Notes in Mathematics

For information about earlier volumes
please contact your bookseller or Springer
LNM Online archive: springerlink.com

Séminaire de Probabilités XXXIV (2000)

Vol. 1730: S. Graf, H. Luschgy, Foundations of Quantization for Probability Distributions (2000)

Vol. 1731: T. Hsu, Quilts: Central Extensions, Braid Actions, and Finite Groups (2000)

Vol. 1732: K. Keller, Invariant Factors, Julia Equivalences and the (Abstract) Mandelbrot Set (2000)

Vol. 1733: K. Ritter, Average-Case Analysis of Numerical Problems (2000)

Vol. 1734: M. Espedal, A. Fasano, A. Mikelić, Filtration in Porous Media and Industrial Applications. Cetraro 1998. Editor: A. Fasano. 2000.

Vol. 1735: D. Yafaev, Scattering Theory: Some Old and New Problems (2000)

Vol. 1736: B. O. Turesson, Nonlinear Potential Theory and Weighted Sobolev Spaces (2000)

Vol. 1737: S. Wakabayashi, Classical Microlocal Analysis in the Space of Hyperfunctions (2000)

Vol. 1738: M. Émery, A. Nemirovski, D. Voiculescu, Lectures on Probability Theory and Statistics (2000)

Vol. 1739: R. Burkard, P. Deuflhard, A. Jameson, J.-L. Lions, G. Strang, Computational Mathematics Driven by Industrial Problems. Martina Franca, 1999. Editors: V. Capasso, H. Engl, J. Periaux (2000)

Vol. 1740: B. Kawohl, O. Pironneau, L. Tartar, J.-P. Zolesio, Optimal Shape Design. Tróia, Portugal 1999. Editors: A. Cellina, A. Ornelas (2000)

Vol. 1741: E. Lombardi, Oscillatory Integrals and Phenomena Beyond all Algebraic Orders (2000)

Vol. 1742: A. Unterberger, Quantization and Non-holomorphic Modular Forms (2000)

Vol. 1743: L. Habermann, Riemannian Metrics of Constant Mass and Moduli Spaces of Conformal Structures (2000)

Vol. 1744: M. Kunze, Non-Smooth Dynamical Systems (2000)

Vol. 1745: V. D. Milman, G. Schechtman (Eds.), Geometric Aspects of Functional Analysis. Israel Seminar 1999-2000 (2000)

Vol. 1746: A. Degtyarev, I. Itenberg, V. Kharlamov, Real Enriques Surfaces (2000)

Vol. 1747: L. W. Christensen, Gorenstein Dimensions (2000)

Vol. 1748: M. Ruzicka, Electrorheological Fluids: Modeling and Mathematical Theory (2001)

Vol. 1749: M. Fuchs, G. Seregin, Variational Methods for Problems from Plasticity Theory and for Generalized Newtonian Fluids (2001)

Vol. 1750: B. Conrad, Grothendieck Duality and Base Change (2001)

Vol. 1751: N. J. Cutland, Loeb Measures in Practice: Recent Advances (2001)

Vol. 1752: Y. V. Nesterenko, P. Philippon, Introduction to Algebraic Independence Theory (2001)

Vol. 1753: A. I. Bobenko, U. Eitner, Painlevé Equations in the Differential Geometry of Surfaces (2001)

Vol. 1754: W. Bertram, The Geometry of Jordan and Lie Structures (2001)

Vol. 1755: J. Azéma, M. Émery, M. Ledoux, M. Yor (Eds.), Séminaire de Probabilités XXXV (2001)

Vol. 1756: P. E. Zhidkov, Korteweg de Vries and Nonlinear Schrödinger Equations: Qualitative Theory (2001)

Vol. 1757: R. R. Phelps, Lectures on Choquet's Theorem (2001)

Vol. 1758: N. Monod, Continuous Bounded Cohomology of Locally Compact Groups (2001)

Vol. 1759: Y. Abe, K. Kopfermann, Toroidal Groups (2001)

Vol. 1760: D. Filipović, Consistency Problems for Heath-Jarrow-Morton Interest Rate Models (2001)

Vol. 1761: C. Adelmann, The Decomposition of Primes in Torsion Point Fields (2001)

Vol. 1762: S. Cerrai, Second Order PDE's in Finite and Infinite Dimension (2001)

Vol. 1763: J.-L. Loday, A. Frabetti, F. Chapoton, F. Goichot, Dialgebras and Related Operads (2001)

Vol. 1764: A. Cannas da Silva, Lectures on Symplectic Geometry (2001)

Vol. 1765: T. Kerler, V. V. Lyubashenko, Non-Semisimple Topological Quantum Field Theories for 3-Manifolds with Corners (2001)

Vol. 1766: H. Hennion, L. Hervé, Limit Theorems for Markov Chains and Stochastic Properties of Dynamical Systems by Quasi-Compactness (2001)

Vol. 1767: J. Xiao, Holomorphic Q Classes (2001)

Vol. 1768: M.J. Pflaum, Analytic and Geometric Study of Stratified Spaces (2001)

Vol. 1769: M. Alberich-Carramiñana, Geometry of the Plane Cremona Maps (2002)

Vol. 1770: H. Gluesing-Luerssen, Linear Delay-Differential Systems with Commensurate Delays: An Algebraic Approach (2002)

Vol. 1771: M. Émery, M. Yor (Eds.), Séminaire de Probabilités 1967-1980. A Selection in Martingale Theory (2002)

Vol. 1772: F. Burstall, D. Ferus, K. Leschke, F. Pedit, U. Pinkall, Conformal Geometry of Surfaces in S^4 (2002)

Vol. 1773: Z. Arad, M. Muzychuk, Standard Integral Table Algebras Generated by a Non-real Element of Small Degree (2002)

Vol. 1774: V. Runde, Lectures on Amenability (2002)

Vol. 1775: W. H. Meeks, A. Ros, H. Rosenberg, The Global Theory of Minimal Surfaces in Flat Spaces. Martina Franca 1999. Editor: G. P. Pirola (2002)

Vol. 1776: K. Behrend, C. Gomez, V. Tarasov, G. Tian, Quantum Comohology. Cetraro 1997. Editors: P. de Bartolomeis, B. Dubrovin, C. Reina (2002)

Vol. 1777: E. García-Río, D. N. Kupeli, R. Vázquez-Lorenzo, Osserman Manifolds in Semi-Riemannian Geometry (2002)

Vol. 1778: H. Kiechle, Theory of K-Loops (2002)

Vol. 1779: I. Chueshov, Monotone Random Systems (2002)

Vol. 1780: J. H. Bruinier, Borcherds Products on O(2,1) and Chern Classes of Heegner Divisors (2002)

Vol. 1781: E. Bolthausen, E. Perkins, A. van der Vaart, Lectures on Probability Theory and Statistics. Ecole d' Eté de Probabilités de Saint-Flour XXIX-1999. Editor: P. Bernard (2002)

Vol. 1782: C.-H. Chu, A. T.-M. Lau, Harmonic Functions on Groups and Fourier Algebras (2002)

Vol. 1783: L. Grüne, Asymptotic Behavior of Dynamical and Control Systems under Perturbation and Discretization (2002)

Vol. 1784: L.H. Eliasson, S. B. Kuksin, S. Marmi, J.-C. Yoccoz, Dynamical Systems and Small Divisors. Cetraro, Italy 1998. Editors: S. Marmi, J.-C. Yoccoz (2002)

Vol. 1785: J. Arias de Reyna, Pointwise Convergence of Fourier Series (2002)

Vol. 1786: S. D. Cutkosky, Monomialization of Morphisms from 3-Folds to Surfaces (2002)

Vol. 1787: S. Caenepeel, G. Militaru, S. Zhu, Frobenius and Separable Functors for Generalized Module Categories and Nonlinear Equations (2002)

Vol. 1788: A. Vasil'ev, Moduli of Families of Curves for Conformal and Quasiconformal Mappings (2002)

Vol. 1789: Y. Sommerhäuser, Yetter-Drinfel'd Hopf algebras over groups of prime order (2002)

Vol. 1790: X. Zhan, Matrix Inequalities (2002)

Vol. 1791: M. Knebusch, D. Zhang, Manis Valuations and Prüfer Extensions I: A new Chapter in Commutative Algebra (2002)

Vol. 1792: D. D. Ang, R. Gorenflo, V. K. Le, D. D. Trong, Moment Theory and Some Inverse Problems in Potential Theory and Heat Conduction (2002)

Vol. 1793: J. Cortés Monforte, Geometric, Control and Numerical Aspects of Nonholonomic Systems (2002)

Vol. 1794: N. Pytheas Fogg, Substitution in Dynamics, Arithmetics and Combinatorics. Editors: V. Berthé, S. Ferenczi, C. Mauduit, A. Siegel (2002)

Vol. 1795: H. Li, Filtered-Graded Transfer in Using Noncommutative Gröbner Bases (2002)

Vol. 1796: J.M. Melenk, hp-Finite Element Methods for Singular Perturbations (2002)

Vol. 1797: B. Schmidt, Characters and Cyclotomic Fields in Finite Geometry (2002)

Vol. 1798: W.M. Oliva, Geometric Mechanics (2002)

Vol. 1799: H. Pajot, Analytic Capacity, Rectifiability, Menger Curvature and the Cauchy Integral (2002)

Vol. 1800: O. Gabber, L. Ramero, Almost Ring Theory (2003)

Vol. 1801: J. Azéma, M. Émery, M. Ledoux, M. Yor (Eds.), Séminaire de Probabilités XXXVI (2003)

Vol. 1802: V. Capasso, E. Merzbach, B.G. Ivanoff, M. Dozzi, R. Dalang, T. Mountford, Topics in Spatial Stochastic Processes. Martina Franca, Italy 2001. Editor: E. Merzbach (2003)

Vol. 1803: G. Dolzmann, Variational Methods for Crystalline Microstructure – Analysis and Computation (2003)

Vol. 1804: I. Cherednik, Ya. Markov, R. Howe, G. Lusztig, Iwahori-Hecke Algebras and their Representation Theory. Martina Franca, Italy 1999. Editors: V. Baldoni, D. Barbasch (2003)

Vol. 1805: F. Cao, Geometric Curve Evolution and Image Processing (2003)

Vol. 1806: H. Broer, I. Hoveijn. G. Lunther, G. Vegter, Bifurcations in Hamiltonian Systems. Computing Singularities by Gröbner Bases (2003)

Vol. 1807: V. D. Milman, G. Schechtman (Eds.), Geometric Aspects of Functional Analysis. Israel Seminar 2000-2002 (2003)

Vol. 1808: W. Schindler, Measures with Symmetry Properties (2003)

Vol. 1809: O. Steinbach, Stability Estimates for Hybrid Coupled Domain Decomposition Methods (2003)

Vol. 1810: J. Wengenroth, Derived Functors in Functional Analysis (2003)

Vol. 1811: J. Stevens, Deformations of Singularities (2003)

Vol. 1812: L. Ambrosio, K. Deckelnick, G. Dziuk, M. Mimura, V. A. Solonnikov, H. M. Soner, Mathematical Aspects of Evolving Interfaces. Madeira, Funchal, Portugal 2000. Editors: P. Colli, J. F. Rodrigues (2003)

Vol. 1813: L. Ambrosio, L.A. Caffarelli, Y. Brenier, G. Buttazzo, C. Villani, Optimal Transportation and its Applications. Martina Franca, Italy 2001. Editors: L. A. Caffarelli, S. Salsa (2003)

Vol. 1814: P. Bank, F. Baudoin, H. Föllmer, L.C.G. Rogers, M. Soner, N. Touzi, Paris-Princeton Lectures on Mathematical Finance 2002 (2003)

Vol. 1815: A. M. Vershik (Ed.), Asymptotic Combinatorics with Applications to Mathematical Physics. St. Petersburg, Russia 2001 (2003)

Vol. 1816: S. Albeverio, W. Schachermayer, M. Talagrand, Lectures on Probability Theory and Statistics. Ecole d'Eté de Probabilités de Saint-Flour XXX-2000. Editor: P. Bernard (2003)

Vol. 1817: E. Koelink, W. Van Assche(Eds.), Orthogonal Polynomials and Special Functions. Leuven 2002 (2003)

Vol. 1818: M. Bildhauer, Convex Variational Problems with Linear, nearly Linear and/or Anisotropic Growth Conditions (2003)

Vol. 1819: D. Masser, Yu. V. Nesterenko, H. P. Schlickewei, W. M. Schmidt, M. Waldschmidt, Diophantine Approximation. Cetraro, Italy 2000. Editors: F. Amoroso, U. Zannier (2003)

Vol. 1820: F. Hiai, H. Kosaki, Means of Hilbert Space Operators (2003)

Vol. 1821: S. Teufel, Adiabatic Perturbation Theory in Quantum Dynamics (2003)

Vol. 1822: S.-N. Chow, R. Conti, R. Johnson, J. Mallet-Paret, R. Nussbaum, Dynamical Systems. Cetraro, Italy 2000. Editors: J. W. Macki, P. Zecca (2003)

Vol. 1823: A. M. Anile, W. Allegretto, C. Ringhofer, Mathematical Problems in Semiconductor Physics. Cetraro, Italy 1998. Editor: A. M. Anile (2003)

Vol. 1824: J. A. Navarro González, J. B. Sancho de Salas, $\mathscr{C}^\infty$ – Differentiable Spaces (2003)

Vol. 1825: J. H. Bramble, A. Cohen, W. Dahmen, Multiscale Problems and Methods in Numerical Simulations. Martina Franca, Italy 2001. Editor: C. Canuto (2003)

Vol. 1826: K. Dohmen, Improved Bonferroni Inequalities via Abstract Tubes. Inequalities and Identities of Inclusion-Exclusion Type. VIII, 113 p, 2003.

Vol. 1827: K. M. Pilgrim, Combinations of Complex Dynamical Systems. IX, 118 p, 2003.

Vol. 1828: D. J. Green, Gröbner Bases and the Computation of Group Cohomology. XII, 138 p, 2003.

Vol. 1829: E. Altman, B. Gaujal, A. Hordijk, Discrete-Event Control of Stochastic Networks: Multimodularity and Regularity. XIV, 313 p, 2003.

Vol. 1830: M. I. Gil', Operator Functions and Localization of Spectra. XIV, 256 p, 2003.

Vol. 1831: A. Connes, J. Cuntz, E. Guentner, N. Higson, J. E. Kaminker, Noncommutative Geometry, Martina Franca, Italy 2002. Editors: S. Doplicher, L. Longo (2004)

Vol. 1832: J. Azéma, M. Émery, M. Ledoux, M. Yor (Eds.), Séminaire de Probabilités XXXVII (2003)

Vol. 1833: D.-Q. Jiang, M. Qian, M.-P. Qian, Mathematical Theory of Nonequilibrium Steady States. On the Frontier of Probability and Dynamical Systems. IX, 280 p, 2004.

Vol. 1834: Yo. Yomdin, G. Comte, Tame Geometry with Application in Smooth Analysis. VIII, 186 p, 2004.

Vol. 1835: O.T. Izhboldin, B. Kahn, N.A. Karpenko, A. Vishik, Geometric Methods in the Algebraic Theory of Quadratic Forms. Summer School, Lens, 2000. Editor: J.-P. Tignol (2004)

Vol. 1836: C. Năstăsescu, F. Van Oystaeyen, Methods of Graded Rings. XIII, 304 p, 2004.

Vol. 1837: S. Tavaré, O. Zeitouni, Lectures on Probability Theory and Statistics. Ecole d'Eté de Probabilités de Saint-Flour XXXI-2001. Editor: J. Picard (2004)

Vol. 1838: A.J. Ganesh, N.W. O'Connell, D.J. Wischik, Big Queues. XII, 254 p, 2004.

Vol. 1839: R. Gohm, Noncommutative Stationary Processes. VIII, 170 p, 2004.

Recent Reprints and New Editions